Edwin A. Heller

Essentials of Materia Medica Pharmacy and Prescription Writing

Edwin A. Heller

Essentials of Materia Medica Pharmacy and Prescription Writing

ISBN/EAN: 9783337177980

Printed in Europe, USA, Canada, Australia, Japan

Cover: Foto ©berggeist007 / pixelio.de

More available books at **www.hansebooks.com**

ESSENTIALS
OF
MATERIA MEDICA

PHARMACY
AND
PRESCRIPTION WRITING

ARRANGED IN CONFORMITY WITH THE CLASSIFICATION IN THE LAST
EDITION OF PROF. H. C. WOOD'S "THERAPEUTICS" AND
FOLLOWING THE COURSE OF PHARMACY AS TAUGHT
IN THE UNIVERSITY OF PENNSYLVANIA

BY

EDWIN A. HELLER, M.D.

QUIZ-MASTER IN MATERIA MEDICA AND PHARMACY AT THE MEDICAL
INSTITUTE, UNIVERSITY OF PENNSYLVANIA

PHILADELPHIA
P. BLAKISTON, SON & CO.
1012 WALNUT STREET
1897

PREFACE.

By the kind permission of Prof. H. C. Wood the author has been enabled to prepare the following volume. The definitions are mostly those given by the last edition of the United States Pharmacopœia. That part of the work relating to pharmacy has been arranged with special reference to the course as given at the University of Pennsylvania. Special attention has been given to the metric system, heretofore too much neglected.

The author trusts that this little work will lighten the labors co-existent with the entrance into the study of materia medica, and if it accomplishes this end he will feel amply repaid for the care and exertion required to compile it.

EDWIN A. HELLER.

934 FRANKLIN STREET, PHILADELPHIA,
July, 1897.

CONTENTS.

PART I.

CHAPTER I.
Definitions—Parts of a Prescription—Kinds of Prescriptions, Superscription, Inscription, ... 8

CHAPTER II.
Apothecaries' Weight and Measure—Domestic Measures—Variation in Size of Drops, ... 17

CHAPTER III.
Metric System—Tables for the Conversion of Apothecaries' Weight and Measure into Metric and vice versa, . 23

CHAPTER IV.
Method of Writing Prescriptions—Conversion of Prescriptions Written in One System into the Other, ... 35

CHAPTER V.
The Grammatical Construction of Prescriptions—Use of Latin: Reasons therefor, Rules thereof—Parts of Speech—Abbreviations, ... 43

CONTENTS.

CHAPTER VI.

Directions to Apothecary—Latin Phrases and Abbreviations—Numerals, ... 61

CHAPTER VII.

Administration: Modes of, Rules for—Doses: Rules for, for Children—Cumulative Action, ... 71

CHAPTER VIII.

Combination of Medicines—Incompatibles, ... 77

PART II.

CHAPTER I.

Materia Medica—Pharmacy—Therapeutics—Officinal and Non-officinal Preparations, ... 85

CHAPTER II.

Average Doses—Tinctures, Extracts, Fluid Extracts, etc., 117

CHAPTER III.

Officinal Drugs and Preparations—Important Non-officinal Drugs—Doses, ... 121

CHAPTER IV.

Poisons—Treatment and Antidotes, ... 359

ADDENDA.

List of Natural Orders, ... 373

INDEX, ... 383

ERRATA.

Page 115, first line, should read, "Solutions of gun-cotton in ether.

Page 131, third line from bottom, "contain" should be "contains."

Page 167, tenth line from bottom, "or" should be "and."

PART I.
PRESCRIPTION WRITING.

CHAPTER I.

DEFINITIONS.—PARTS OF A PRESCRIPTION.

The word prescription (from the Latin *præ*, before, and *scriptum*, perfect participle of *scribo*, to write, meaning, written,), was at one time understood to mean any direction whatever, either verbal or written, given to the patient. It included directions as to diet, ventilation, heat, light of his apartment, etc.; in fact, any direction whatever relating to the care of the patient or his surroundings.

At present, however, a prescription is generally understood to be a written formula containing the names and quantities of a drug or drugs, together with directions to the apothecary for dispensing, and directing also the patient as to the manner, method, and frequency of administration.

A prescription may be either (1) simple or (2) compound; and the formula it contains may be

either (1) officinal or (2) extemporaneous or magistral.

A simple prescription contains but one ingredient.

Compound prescriptions always contain more than one ingredient.

An **officinal preparation** or formula is one published and authorized by the United States Pharmacopeia, and although it may contain numerous ingredients, in prescribing it is necessary to write only the officinal name, together with the dose and directions.

An **extemporaneous** or **magistral formula** is one composed by the physician to suit the individual case.

A typical prescription consists of:

1. The **superscription**, heading.
2. The **inscription**, names and quantities of the drugs prescribed.
3. The **subscription**, directions to the pharmacist.
4. The **signature**, directions to patient.
5. **Date, and the signature** of the physician.

I. In English the *superscription* is always the symbol ℞; a combination of R from *recipe* (Latin, imperative of *recipio*), take, and the zodiac sign ♃. Originally, prescriptions were always begun with an invocation to Jove or Jupiter, and his blessing invoked on the action of the remedy, whence we derive

the combination of the R and ♃. When Christianity supplanted the heathen beliefs, the prayers were abbreviated and changed in various ways: as, A, Ω, the beginning and end, the first and last, the alpha and omega of everything; JJ, Juvante Jesu (Jesus helping); JD, Juvante Deo (God helping); ND, Nomine Dei (in God's name); and at one time also the simple +, the sign of the cross. But all these have been discarded for the old ℞. In France P or Ps (*prenez*, take,) is employed.

II. The *inscription*, or body of the prescription, contains the names and quantities of the ingredients, and in a typical prescription includes:

1. The **basis**, the principal active agent.
2. The **adjuvant**, or auxiliary intended to aid and increase the action of the basis.
3. The **corrective** or **corrigent**, to correct or modify one or both of the above two.
4. The **vehicle** or **excipient**, to render palatable, assimilable, or easy of administration the entire prescription.

The old maxim of Asclepiades, "*Curare cite tute et jucunde*," might be applied as follows:

Curare	(cure),	with the	*Basis*.
Cite	(quickly),	" "	*Adjuvant*.
Tute	(safely),	" "	*Corrective*.
et Jucunde	(pleasantly),	" "	*Vehicle*.

The names of the ingredients are always written in

Latin and are in the *genitive case*, governed by *recipe*.

The quantities, if written out in Latin, which, however, is practically never done, are always in the *accusative*.

The *subscription* or instruction as to method of dispensing is always in Latin. (A list of the principal phrases will be found on pages 61–65.)

The *signature*, written Signa, or Sig., consists of the directions to the patient, is always in English, and should always be as simple and distinct as it possibly can be written. Even in cases where it is desired to conceal from the pharmacist the purpose for which the remedy is intended, this can be done at no sacrifice of meaning; *e. g.*, in case an injection should be ordered we may simply sign, "Use as wash;" "Bathe affected part two or three times daily," etc. Thus the druggist is often unable to tell if the prescription is intended for eye, mouth, vagina, urethra, or other part of the body, and the patient protected both while procuring his prescription and at home, should the preparation be seen by those who have no right to do so.

CHAPTER II.

APOTHECARIES' WEIGHT AND MEASURE.

At present, although the metric system is really the only system founded on a definite scientific basis, both the **apothecaries'** weight and measure and the metric system are employed.

APOTHECARIES' WEIGHT.

20 grains (gr.) = 1 scruple.
60 grains or 3 scruples (Ɖ) = 1 dram.
480 grains or 8 drams (ʒ) = 1 ounce.
5760 grains or 96 drams or 12 ounces (℥) = 1 pound.

In Latin, respectively:

Pound, symbol ℔, = libra.
Ounce, " ℥, = uncia.
Dram, " ʒ, = drachma.
Scruple, " Ɖ, = scrupulum.
Grain, " gr., = granum.

The scruple (Ɖ) is practically obsolete, because if not carefully written it is easily confounded with the dram (ʒ), and thus may give rise to serious consequences. Amounts less than one dram should be expressed in *grains*.

The British pharmacopeia directs that the pound should contain 16 ounces, each of them equaling 437.5 grains, or 7000 grains to the pound, being

considerably more than our pound. This must be borne in mind in the use of formulæ based on the British standard.

APOTHECARIES' OR WINE MEASURE.

60 minims (♏) = 1 fluidram.
8 fluidrams (f ʒ) = 1 fluidounce, or 8 f ʒ, or 480 ♏.
16 fluidounces (f ℥) = 1 pint, or 16 f ℥, or 128 f ʒ, or 7680 ♏.
8 pints (O) = 1 gallon (C), or 8 O, or 128 f ℥, or 1024 f ʒ, or 61,440 ♏.

In Latin, respectively:

The gallon, symbol C, = congius.
" pint, " O, = octarius.
" fluidounce, " f ℥, = fluiduncia.
" fluidram, " f ʒ, = fluidrachma.
" minim, " ♏, = minimum.

The English *pint* contains 20 fluidounces, and the fluidounce equals 7 fluidrams and 2½ minims; their *minim*, therefore, is equal only to .96 of ours, theirs weighing approximately .91 of a grain, while ours is equivalent to .95 of a grain.

As practically all patients are unfamiliar with apothecaries' measure, we must employ some **domestic** measure fairly equivalent to it, and for that purpose we generally consider—

The drop = a minim.
" teaspoon = a fluidram.
" dessertspoon = 2 fluidrams.
" tablespoon = 4 fluidrams, or ½ of a fluidounce.
" wineglass = 2 fluidounces.
" teacup = 4 fluidounces.

Naturally, it is at once evident that as the drops of some liquids are much larger than those of others, and as teaspoons vary greatly in capacity, to say nothing of the balance, the method is faulty and not to be relied upon for accurate dosage.

In such cases it is necessary to procure an accurate graduate at once and use only this in giving the medicine to the patient.

If drops are ordered (as for use in eye work), we may order a pipet. Should we desire to give only minims in this manner, accurate minim pipets may be used. That the size of drops varies greatly may easily be seen by glancing at the following list of drugs. Drops in one fluidram:

Dilute sulphuric acid	= 54 to 48
Aromatic sulphuric acid	= 115 " 145
Hydrocyanic acid	= 45 " 48
Ether	= 150 " 160
Chloroform	= 180 " 275
Alcohol	= 120 " 143
Castor oil	= 50 " 58
Tinct. aconite	= 115 " 125
Tinct. opium	= 105 " 145

CHAPTER III.

METRIC SYSTEM.

The **metric system** now used practically exclusively in France and Germany is being rapidly adopted in the other countries of Europe, and is making rapid progress, as it deserves to, in this country. It is the only rational system of weights and measures we possess, the unit of length being the **meter**, which equals one forty-millionth of the earth's circumference through the poles, or one ten-millionth of the distance from the pole to the equator. It equals 39.37 inches, being 3⅓ inches (about) more than our yard.

The **gram**, the unit of weight, is the weight of one cubic centimeter (c. c.) of water at its greatest density (4° C. or 39° F.).

In writing a prescription according to the metric system, if we desire all the ingredients to be weighed, we merely place the symbol gm. above the figures. If, however, the liquids are to be measured, we write gm. and c. c. (cubic centimeters).

The gram (solid) equals 15.432 grains. The gram (of water) measures 16.231 minims.

The subdivisions of the units are formed by prefixing to the unit the Latin—

  Milli (from *mille*)   = $\frac{1}{1000}$ of the unit.
  Centi (from *centum*) = $\frac{1}{100}$ of the unit.
  Deci (from *decem*) = $\frac{1}{10}$ of the unit.

Thus:

  Two decigrams   = $\frac{2}{10}$ of a gram.
  Or two millimeters = $\frac{2}{1000}$ of a meter.
  Or one centiliter   = $\frac{1}{100}$ of a liter.

The multiples of the units are formed by prefixing to the unit the Greek—

  Deca   (from Δεκα) = 10 times the unit.
  Hecto (from Ἑκατον) = 100 times the unit.
  Kilo   (from Κίλιος) = 1000 times the unit.
  Myria (from Μυριας) = 10,000 times the unit.

Thus:

  A decaliter   = 10 liters.
  A hectometer = 100 meters.
  A kilogram   = 1000 grams, etc.

Thus the multiple and subdivisions would be of the—

Gram.		*Meter.*		*Liter.*
Milligram,	$\frac{1}{1000}$	millimeter,	$\frac{1}{1000}$	milliliter.
Centigram,	$\frac{1}{100}$	centimeter,	$\frac{1}{100}$	centiliter.
Decigram,	$\frac{1}{10}$	decimeter,	$\frac{1}{10}$	deciliter.
Gram,		meter,		liter.
Decagram,	10	decameters,	10	decaliters.
Hectogram,	100	hectometers,	100	hectoliters.
Kilogram,	1000	kilometers,	1000	kiloliters.

 1 gram, . . 15.432 grains. 1 meter, . . 39.37 inches.
     1 liter, . . 2.113 pints.

Instead, however, of writing out in full the prefixes of the units, we employ the **decimal system** entirely; thus:

2	or	2
.125	"	125
.025	"	025

The whole numbers always signify grams and cubic centimeters, according to the symbol at the top; thus: gm. and c. c. If, however, there is no symbol, then all the ingredients—liquid and solid—are supposed to be weighed out in grams. Thus the above amounts would equal 2 grams; $\frac{125}{1000}$ of a gram or 12½ centigrams; $\frac{25}{1000}$ of a gram or 2½ centigrams or 25 milligrams.

Although it is sufficient to merely indicate the decimals by a point, as in the first example, in writing prescriptions it is policy to always use the line, as shown in the second example, so as to leave no possible room for doubt, as a spot in the paper may be mistaken for the point, and this would, of course, multiply or divide the result, possibly ten, possibly a hundred, fold.

For all practical purposes, and for converting apothecaries' weight into the metric, and *vice versâ*, a gram may be considered equal to 15 or 16 grains, using that number which divides or multiplies most easily. Except in cases of poisons, alkaloids, and very powerful drugs, where it is well to consider that 15⅔ grains equal 1 gram.

Likewise, with the liquid preparations, either 15 or 16 minims may be considered equivalent to one cubic centimeter.

The fluidram may be said to equal 4 c. c. The dram to equal 4 grams. The fluidounce may be considered as equivalent to 30 or 32 c. c.; and the ounce, 30 or 32 grams. The liter is equal to 2.113 pints.

It will be seen that the table for liquids is computed for water at 4° C.; and, consequently, to be strictly accurate, allowance would have to be made for the weight of all the liquid preparations compared to that of water. Thus, the dose of all liquids lighter than water: *e. g.*, alcohol, ether, the tinctures, etc., would be slightly less; the dose of heavier liquids: *e. g.*, the syrups, glycerites, decoctions, etc., would be larger in order to be absolutely correct. But practically the difference is so small that in most, if not all, cases it may be discarded.

TABLE TO CONVERT METRIC INTO APOTHECARIES' WEIGHT, AND *Vice Versâ*.

	1 grain	=	.06 gram.
℈j =	20 grains	=	1.2 grams.
ʒj =	60 "	=	4. "
℥j =	480 "	=	30. or 32 grams.
	1 ♏	=	.06 c. c.
fʒj =	60 ♏	=	4. "
f℥j =	480 ♏	=	30. or 32 c. c.

It will be seen that the figures do not accord ex-

actly with the results of multiplication, but we even them up in order to obviate the fractions.

    1.   gram = 15. or 16 grains.
    .1   "   = 1.5 grains.
   .01   "   = .15 grain.

    1.   c. c. = 16. ♏.
    .1   c. c. = 1.6 ♏.
   .01 c. c. = .16 ♏.

ESSENTIALS OF PRESCRIPTION WRITING. 33

TABLE FOR CONVERTING APOTHECARIES' WEIGHT INTO THE METRIC SYSTEM, AND *Vice Versâ.*

SOLIDS.		LIQUIDS.	
Apothecaries'.	Metric.	Apothecaries'.	Metric.
GRAINS.	GRAMS.	MINIMS.	GRMS AND C.C.
$\frac{1}{64}$	.001	1	.06
$\frac{1}{40}$	.0015	2	.12
$\frac{1}{30}$	.002	3	.18
$\frac{1}{20}$	.003	4	.24
$\frac{1}{18}$	.004	5	.3
$\frac{1}{12}$	.005	6	.36
$\frac{1}{10}$	.006	7	.42
$\frac{1}{8}$	.008	8	.5
$\frac{1}{4}$	.016	9	.55
$\frac{1}{3}$	.02	10	.6
$\frac{1}{2}$	.03	12	.72
1	.065	16	1.
2	.013	20	1.25
3	.2	25	1.55
4	.26	48	3.
5	.32	50	3.12
10	.65		
15	1.	(f ℥) 60	3.75
(Ə) 20	1.3	240	15.
30	1.95	(f ℥) 480	30.
(℥) 60	3.75		
(℥) 480	30.		

CHAPTER IV.

METHOD OF WRITING PRESCRIPTIONS.—CONVERSION OF APOTHECARIES' WEIGHT AND MEASURE INTO THE METRIC SYSTEM, AND VICE VERSA.

The proper, as well as the quickest and safest, way to write a prescription is to put down the names of the ingredients intended to be used; then determine the number of doses to be given *in toto;* and, finally, after multiplying the individual dose of each ingredient by the number of doses, put the corresponding amount opposite each drug.

Example: If we wish to write for the compound cathartic pill of the United States Pharmacopeia:

		Gm.
℞.	Extracti colocynthidis compositi, ...	8.
	Hydrargyri chloridi mitis, ...	6.
	Extracti jalapæ, ...	3.
	Cambogiæ, ...	15.
	Aquæ, quantum sufficiat.	
M.	Ft. pilulæ No. x.	

Having decided to give 10 pills, and deciding the single dose of colocynth to be 8 centigrams, we multiply 8 centigrams by 10, equaling 8 decigrams, which we put opposite the colocynth.

The single dose of calomel we are going to give

being 6 centigrams, we multiply this by 10, equaling 6 decigrams, which we then place opposite its line, and so on.

The quantities of the drugs in metric prescriptions are expressed always in the Arabic numerals; while in the apothecaries' weight we use the Roman numerals, except in the case of fractions, where, for greater accuracy, we use ordinary figures; or in cases where a large or ordinarily poisonous dose is intended we may place the Arabic numeral in parenthesis alongside the Roman, in order to assure the druggist that a large amount is intended; thus f℥iij (3).

For converting apothecaries' weight into the metric, or *vice versâ*, reference to the table at the end of chapter III will obviate the necessity of multiplication and division. It is, however, advisable, for the sake of practice, that the student convert several prescriptions without the table, in order to be familiar with the method and able at any time to convert one table into the other. For example, to convert the following into the metric system:

For JOHN SMITH.

℞. Pulveris extracti glycyrrhizæ,
    " acaciæ, āā gr. viij.
    " sacchari, gr. x.
    " kino, gr. ij.
    Spiritus aromatici, ♏v.
    Mellis despumati, q. s.

M. Ft. massa, in pilulas numero triginta dividenda.

SIGNA.—One after meals.

601 LANCET AVE., *Nov. 15, 1896.* JOHN MEDICUS, M.D.

Considering 1 gram to equal 16 grains (see chap. III) we find the first and second ingredients to equal ½ of a gram, which we would write | 5. The third equals $\frac{10}{16} = \frac{5}{8}$ or | 625 grams or 62½ centigrams.

The fourth equals $\frac{2}{16} = \frac{1}{8}$ or | 125 grams or 12½ centigrams. The last quantity, 5 ♏, would be expressed in cubic centimeters, and as we may count either 15 or 16 minims as equaling 1 c. c., we will take 15; therefore, we would have $\frac{5}{15}$ or ⅓ of a cubic centimeter or 3⅓ cubic decometers, written | 33.

The body of the prescription then would read:

			gm.	c. c.
℞.	Pulveris extracti glycyrrhizæ,			
	" acaciæ,	aa	5	
	" sacchari,		625	
	" kino,		125	
	Spiritus aromatici,			33
	Mellis despumati, q. s.			

Or, *vice versâ*, to convert the following into apothecaries' weight:

For Sam Small.

℞.	Potassii bromidi,	15	
	Antipyrin,	7	5
	Acidi arsenosi,		06
M.	Fiant pulveres numero viginti.		
Signa.—One at bedtime.			

1428 Edgely St., *Nov. 15, 1896.*   Thos. Jones, M.D.

Now, 1 gram equals 15 or 16 grains; 15 (grams) × 16 (grains) = 240; 240 grains ÷ 60 (the number of grains in a dram) = 4; consequently the total is ℥iv.

The second ingredient calls for 7.5 grams: $7\frac{1}{2}$ (grams) $\times$ 16 (grains) = 120; 120 (grains) ÷ 60 (grains in a dram) = 2, hence ℨij. The third quantity: .06 grams or 6 centigrams = $\frac{6}{100}$ grams $\times$ 16 (the number of grains in a gram) = $\frac{96}{100}$; practically, 1 grain. Of course, there is a difference of $\frac{4}{100}$ grains, but as there are 20 doses, the difference in each dose would be but $\frac{4}{2000}$ of a grain, which is insignificant and may be disregarded. In fact, it is the rule in transposing from one system to another, to always even up the amounts: unless the drug is extremely potent, it will be found that the difference one way or the other will be too small to be of practical import. And the prescription then would read:

For SAM SMALL.

℞. Potassii bromidi, ℨiv.
    Antipyrin, ℨij.
    Acidi arsenosi, gr. j.
M. Fiant pulveres numero viginti.
SIGNA.—One at bedtime.
1428 EDGELY ST., *Nov. 15, 1896.*   THOS. JONES, M.D.

CHAPTER V.

THE GRAMMATICAL CONSTRUCTION OF PRESCRIPTIONS.

Latin is the language *par excellence* for prescriptions. Although the physician may use English or any other language if he sees fit, the arguments in favor of Latin far outweigh any which may be brought against it.

First, it is a "dead" language, does not undergo any change, and words expressed in Latin are understood all over the civilized world, whereas if we wrote prescriptions only in the current tongue, special knowledge of that language would be necessary to translate it into any other language. The comprehension, however, of even a very moderate amount of Latin enables us to understand a prescription written in any of the civilized countries. Latin is the universal language of science,—the botanic and chemic names of all our remedies are in Latin.

When we express the name of a drug in Latin it refers distinctly and positively to only one drug, whereas the English word may include a number of drugs entirely different from one another. Thus,

cimicifuga means only the cimicifuga racemosa, whereas the English name "snakeroot" is applied to numerous plants, each differing according to its habitat. If we write Indian hemp, either apocynum cannabinum or cannabis indica may be dispensed. Wintergreen may mean either chimaphila or gaultheria.

Checkerberry may mean either uva ursi or gaultheria.

In any or all such cases the drug dispensed may have directly opposite effect on the patient from that desired, and not alone may it be inert, but it may be positively harmful, if not disastrous.

Again, there is a certain prejudice against the use of certain drugs,—as mercury and iodid of potassium,—and should the prescription be in English the patient may refuse to take it.

Again, it is sometimes desirable that a patient should not know the exact nature of the drugs he is taking, or he may wish to take our prescription to a foreign country. In either case Latin fulfils all requirements.

The directions to the patient, however, should invariably be written in English. There is absolutely no valid reason why he should not read them; if they be written in Latin, should the patient forget the instructions—and patients are often singularly negligent in this respect—he would be at a loss how to take the remedy, and at what times.

ESSENTIALS OF PRESCRIPTION WRITING.

A certain rudimentary knowledge of Latin is necessary in order to write a proper prescription.

The following few simple rules are indispensable:

1. The names of the drugs are always in the genitive if the quantity is expressed (governed by the ℞ (take) amount *of*).

2. If no quantity is expressed, but a numeral adjective follows, the noun is always in the accusative.

3. The quantities are always in the accusative, governed by the imperative *recipe* (take).

4. Adjectives agree with their nouns in number, gender, and case.

Verbs, prepositions, conjunctions, and **adverbs** undergo no change, the principal ones being:

Verbs.	*Abbreviations.*	*English.*
Recipe,	℞,	take.
Misce,	M.,	mix, or mix thou.
Signa,	Sig.,	mark or label.
Fiat,	Ft.,	let (it) be made.
Fiant,	Ft.,	let (them) be made.
Adde,	Add.,	add.
Bulliat,	Bull.,	let boil.
Cola,	Col.,	strain.
Divide,	Div.,	divide.
Macera,	Mac.,	macerate.
Repetatur,	Rep.,	repeat.
Solve,	Sol.,	dissolve.
Sufficit,	Suf.,	it suffices.
Tere,	Ter.,	rub.

Prepositions, etc.

Cum,	with.
In,	in, into.
Ad,	to, up to.
Et,	and.
Ana, ā̄ā,	of each.

FIRST DECLENSION.

All pharmacopeial nouns ending in a belong to the first declension, are feminine gender, and are declined as follows, except **aspidosperma, physostigma**, and **folia** (plural):

Singular.	*Plural.*
Nom., gutt*a* (a drop), —a.	Gutt*æ* (drops), —æ.
Gen., gutt*æ* (of a drop), —æ.	Gutt*arum* (of drops), —arum.
Acc., gutt*am* (a drop), —am.	Gutt*as* (drops), —as.

The stem (*gutt*) remains unchanged, and if the case-endings are committed to memory there will be no trouble in declining any noun.

(**Aspidosperma** and **physostigma** change to **aspidospermatis** and **physostigmatis** in the genitive. **Folia** (leaves) is plural; genitive, **foliorum**.)

Two pharmacopeial nouns of the first declension end in **e**; *e.g.*, **aloe** and **mastiche**. The case-endings in the singular are: Nom., *e*; gen., *es*; acc., *en*. The plural is like that of *gutta*.

SECOND DECLENSION.

All pharmacopeial nouns ending in **us** (except **Rhus spiritus, quercus,** and **fructus**—fourth

ESSENTIALS OF PRESCRIPTION WRITING. 51

declension) belong to the second declension. They are nearly all masculine, declined thus:

Singular.	Plural.
Nom., syrup*us* (a syrup), –us.	Syrup*i* (syrups), –i.
Gen., syrup*i* (of a syrup), –i.	Syrup*orum* (of syrups), –orum.
Acc., syrup*um* (a syrup), –um.	Syrup*os* (syrups), –os.

All pharmacopeial nouns ending in **um** are neuter, of second declension, and are declined thus:

Singular.	Plural.
Nom., acet*um* (a vinegar), –um.	Acet*a* (vinegars), –a.
Gen., acet*i* (of a vinegar), –i.	Acet*orum* (of vinegars), –orum.
Acc., acet*um* (a vinegar), –um.	Acet*a* (vinegars), –a.

Rhus changes to **Rhois**.

THIRD DECLENSION.

All declinable pharmacopeial nouns whose nominative ends otherwise than in **a, us,** and **um,** are (except **aloes, mastiche, eriodictyon, hæmatoxylon,** and **toxicodendron**) of the third declension. They are declined like **liquor. Digitalis,** however, does not change the genitive. Nouns ending in **as** change to **atis**; as, **acetas, acetatis**;* **carbonas, carbonatis; sulphas, sulphatis,** etc.

Singular.	Plural.
Nom., liquor (a solution).	Liquor*es* (solution), –es.
Gen., liquor*is* (of a solution), –is.	Liquor*um* (of solutions), –um.
Acc., liquor*em* (a solution), –em.	Liquor*es* (of solutions), –es.

---

* See note at end of chapter.

ESSENTIALS OF PRESCRIPTION WRITING. 53

EXCEPTIONS.

Nom.,	anthemis,	Gen.,	anthemidis.
"	cortex,	"	corticis.
"	pepo,	"	peponis.
"	phosphis,	"	phosphitis.
"	sulphis,	"	sulphitis.
"	mucilago,	"	mucilaginis.
"	solidago,	"	solidaginis.
"	colocynth,	"	colocynthidis.
"	hæmatoxylon,	"	hæmatoxyli.
"	radix,	"	radicis.
"	semen,	"	seminis.

FOURTH DECLENSION.

We employ only three nouns of the fourth declension; they are: **Fructus, spiritus** (masculine), and **quercus** (feminine), which are declined thus:

Singular.
Nom., spirit*us* (a spirit), –us.
Gen., spirit*us* (of a spirit), –us.
Acc., spirit*um* (a spirit), –um.

Plural.
Spirit*us* (spirits), –us.
Spirit*uum* (of spirits), –uum.
Spirit*us* (spirits), –us.

The ablative case is used only after **cum** (with); as:

Cum creta (with chalk).
" calce (with lime).
" ferro (with iron).
" semisse (with a half), etc.

The following nouns are indeclinable. The genitive does not change:

Amyl.	Elixir.	Pyrogallol.
Azedarach.	Eucalyptol.	Quebracho.
Buchu.	Hydronaphthol.	Salol.
Cajuputi.	Kino.	Sassafras.
Catechu.	Matico.	Sumbul.
Cusso.	Menthol.	Thymol.
Diachylon.	Naphthol.	Ichthyol.
Digitalis.	Phenol.	Iodol, etc.

Although we practically always express the quantities by means of the **Roman numerals**, should they be written out they are always in the accusative, formed as follows:

1. Nouns of quantity ending in **a** are feminine and have the accusative singular end in **am**; plural, **as**; thus, *drachma*, accusative singular *drachmam;* plural, *drachmas*. *Uncia*, accusative singular *unciam;* plural, *uncias*.

2. Those ending in **us** or **um** have the accusative singular end in **um**; accusative plural of **us** is **os**, and of **um** is **a**; thus:

	Accusative singular.	*Accusative plural.*
Congius,	Congium,	Congios.
Minimum,	Minimum,	Minima.

Adjectives are declined like nouns and agree with them in gender, number, and case.

If the adjective is of the same declension as the

Ammonii chloridum
Mistura glycyrrhizae co
et Sig.

noun, naturally the genitive will be the same. If, however, it is of a different declension, its genitive must be formed according to that declension.
'To illustrate: Suppose we desire to change the following recipe into unabbreviated Latin—

Take of chlorid of ammonium, three drams;
" " compound mixture of licorice, two fluidounces.
Mix, and label teaspoonful three times a day.

THOS. LEIDY, M. D.
For JOHN OWENS, *Nov. 15, 1896.*

For "take of" we write ℞. In Latin, chlorid of ammonium is termed *ammonii chloridum;* but the directions are to take three drams *of* chlorid of ammonium. Consequently, it must be written in the genitive, *ammonii chloridi.* The accusative plural of *dram* is *drachmas;* the accusative plural of three is *tres;* hence, *drachmas tres.* Compound mixture of licorice in Latin is termed *mistura glycyrrhizæ composita;* but the direction states *take of* the compound mixture of licorice. We already have *of licorice* in the phrase; hence, *glycyrrhizæ* is already in the genitive, but *mistura* and *composita* are not; they are, according to our table, of the first declension, and consequently form the genitive by changing a to æ; hence, *misturæ glycyrrhizæ compositæ.* Then the accusative plural of *uncia* would be *uncias;* that of *duo, duas*—*fluiduncias duas.* For mix we write *misce;* for label,

signa; for three times a day, *t. i. d.* Therefore, in Latin the prescription would read :

℞ . Ammonii chloridi, drachmas tres.
    Misturæ glycyrrhizæ compositæ, fluiduncias duas.
Misce. Signa.—Teaspoonful, t. i. d.
                                  THOS. LEIDY, M. D.
For JOHN OWENS, *Nov. 15, 1896.*

NOTE.—The salts of the metals generally ending in as change to atis in the genitive, thus :

Nominative.	*Genitive.*
Acetas,	Acetatis.
Bicarbonas,	Bicarbonatis.
Carbonas,	Carbonatis.
Citras,	Citratis.
Phosphas,	Phosphatis.
Sulphas,	Sulphatis.
Tartras,	Tartratis, etc.

CHAPTER VI.

DIRECTIONS TO THE APOTHECARY.—LATIN PHRASES AND THEIR ABBREVIATIONS.—NUMERALS.

As we desire to give brief *directions to the apothecary*, there have come into use a number of Latin phrases and words which, however, are generally abbreviated as follows:

Latin.	*Abbreviations.*	*English.*
Acidum,	acid.,	an acid.
Ad,	ad,	to, up to.
Ad libitum,	ad lib.,	at pleasure.
Adde,	add.,	add (thou).
Ana,	A., aa,	of each.
Aqua bulliens,	aq. bul.,	water, boiling.
Aqua destillata,	aq. dest.,	water, distilled.
Bene,		well.
Bis in dies,	bis in d.,	twice a day.
Bulliat,	bull.,	boil.
Cape, capiat,	cap.,	take.
Capsula,	caps.,	capsule.
Ceratum,	cerat.,	a cerate.
Chartula,	chart.,	a paper for powder.
Cochleare magnum,	coch. mag.,	tablespoon.
Cochleare parvum,	coch. parv.,	teaspoon.
Cola,	col.,	strain.
Collyrium,	collyr.,	eye-wash.

ESSENTIALS OF PRESCRIPTION WRITING.   63

Latin.	Abbreviations.	English.
Collutorium,	collut.,	mouth-wash.
Compositus,	comp., co.,	compound.
Congius,	C,	gallon.
Confectio,	conf.,	confection.
Cortex,	cort.,	bark.
Cum,		with.
Decoctum,	decoc.,	a decoction.
Dimidius,	dim,	one-half.
Divide,	d., div.,	divide (thou).
Dividendus,	dividend.,	to be divided.
Dividatur in partes æquales,	d. in p. æ.,	divide in equal parts.
Emplastrum,	emp.,	a plaster.
Extende supra,	ex. sup.,	spread upon.
Fac, fiat,	F.,	make, let be made.
Fiant,	F.,	let them be made.
Filtrum, filtra,	fil.,	a filter, filter (v).
Gargarisma,	garg.,	a gargle.
Gutta, guttæ,	gtt.,	drop, drops.
Guttatim,	guttat.,	drop by drop.
Haustus,	haust.,	a draught.
Hora,	h., hor,	an hour.
In dies,	in d.,	daily.
Instar,	inst.,	like (used with genitive).
Lac,		milk.
Lagena,		a bottle.
Libra,	℔., Lb.,	pound (Troy).
Liquor,	liq.,	solution.
Lotio,	lot.,	lotion.
Massa,	mass.,	pill mass.
Mica panis,	mic. pan.,	crumb of bread.
Misce,	M.,	mix.
Mucilago,	mucil.,	a mucilage.

Latin.	Abbreviations.	English.
Numerus, numero,	No.,	number, in number.
Octarius,	O,	pint.
Ovum, ovi,	ov.,	egg.
Pars,	par.,	a part of.
Partes æquales,	p. æ.,	equal parts.
Pediluvium,		foot-bath.
Per fistulam vitream,		through a glass tube.
Phila,	phil.,	a vial.
Pilula,	pil.,	a pill.
Pro re nata,	p. r. n.,	as required.
Pulvis,	pulv.,	a powder.
Quantum sufficiat,	q. s.,	sufficient quantity of (use genitive after).
Quâquâ horâ,	q. h.,	every hour.
Saturatus,	sat.,	saturated.
Scatula,	scat.,	a box.
Semissis, semisse,	ss.,	one-half.
Sesuncia,	sesunc.,	an ounce and a half.
Signa,	S., Sig.,	sign, label.
Sine,		without.
Solve,	solv.,	dissolve.
Statim,	stat.,	at once.
Talis,	tal.,	such.
Tritura,	trit.,	triturate.
Tere simul,	ter. sim.,	rub together.
Tere exactissime,	ter. exact.,	rub finely.
Ter in die,	t. i. d.,	three times a day.
Vehiculum,	vehic.,	vehicle.
Vitellus,	vit,	yolk of egg.

Although it is perfectly allowable to use the above abbreviations, the names of drugs should never be abbreviated if there is any likelihood of one drug

being taken for another. In fact, it is better always to write out the names of the drugs in full, as the many disastrous results which have already been caused by abbreviating have no need of increase.

Thus, **acid. hydroc.**, may mean either hydrochloric or hydrocyanic acid; **ext. col.**, may mean extractum colchici or extractum colocynthidis; **hydra. chlor.**, either chloral hydrate or chlorid of mercury (hydrargyrum), etc.

NUMERALS

Are all indeclinable except *unus, duo, tres, mille*, and the hundreds (100, 200, etc.).

1, unus.	20, viginti.
2, duo.	30, triginta.
3, tres.	40, quadraginta.
4, quatuor.	50, quinquaginta.
5, quinque.	60, sexaginta.
6, sex.	70, septuaginta.
7, septem.	80, octoginta.
8, octo.	90, nonaginta.
9, novem.	100, centum.
10, decem.	200, ducenti.
11, undecim.	300, trecenti.
12, duodecim.	400, quadringenti.
13, tredecim	500, quingenti.
14, quatuordecim.	600, sexcenti.
15, quindecim.	700, septingenti.
16, sexdecim.	800, octingenti.
17, septendecim.	900, nongenti.
18, octodecim.	1,000, mille.
19, novendecim.	10,000, decemmille.

Unum, duo, and tres are thus declined :

Masculine.	Feminine.	Neuter.
Nom., unus,	una,	unum.
Gen., unius,	unius,	unius.
Acc., unum,	unam,	unum.
Nom., duo,	duæ,	duo.
Gen, duorum,	duarum,	duorum.
Acc., duos,	duas,	duo.
Nom., tres,	tres,	tria.
Gen, trium,	trium,	trium.
Acc., tres,	tres,	tria.

CHAPTER VII.

ADMINISTRATION.

Medicines may be given :

1. By **enema** (per rectum, dose about one-third larger than by the mouth).

2. By **hypodermic injection** (subcutaneously, dose about one-third less than by the mouth).

3. By the **mouth** (this method employs the officinal doses).

4. **Epidermically** (by means of friction of the skin after local application).

5. **Endermically** (after denudation of the skin by a blister).

6. **Enepidermically** (direct application to skin).

7. **Insufflation** and **vaporization** (usually to respiratory tract only).

8. **Intravenously** (rarely used).

As a rule, the **doses** stated are intended for adults. In order to employ the same remedies for children several rules have been proposed, viz. :

1. *De Young's Rule.*—Divide the age by the age +12, and divide the adult dose by the result.

2. *Cowling's Rule.*—Divide the age at next birthday by 24.

3. *Clark's Rule.*—This rule assumes the average weight of a normal adult to be about 150 pounds. To find the dose for a person weighing less than that, divide the dose by the weight divided by 150. Thus: For a child weighing 25 pounds, the dose would be $\frac{25}{150}$, or ⅙ of the adult dose. This method, although on a more scientific base than the others, is not often used.

To illustrate De Young's and Cowling's rules: Supposing the adult dose of a drug to be one dram; for a child three years of age we would divide 3 by $3 + 12 = \frac{3}{15}$ or ⅕; hence, this dose would be 12 grains. According to Cowling's rule, we divide 4 by $24 = ⅙$, which equals 10 grains in the above case.

These results are accurate enough for use in all those drugs which are normally well borne by children; it must be borne in mind, however, that some drugs are very poorly borne by children,—for example, morphine, etc.,—while of others,—for example, mercury,—children will often bear well amounts disproportionate to their age and weight.

Doses will also be modified by the general nutrition and vital resistance of the patient, sex (males bear larger doses than females), previous habits (alcoholic, opium, cocain, etc.), previous mode of living, idiosyncrasies, etc.

Some drugs may be taken for long periods of time, but with others the system becomes tolerant and

FOR THERAPEUTIC NOTES.

the dose must be increased to have the same, if any, effect.

Again, other drugs—for example, digitalis—have what is called a cumulative effect; *i.e.*, when given for too long a period of time the system seems to store up the drug and there may suddenly be a poisonous explosion, so to speak, in which the poisonous character of the drug may become disagreeably or even fatally manifest.

FOR THERAPEUTIC NOTES.

CHAPTER VIII.

COMBINATION OF MEDICINES.

Different medicines are combined in one mixture in order to (1) increase the action of the base; (2) correct any undesirable action of the base; (3) to meet more than one indication; (4) to obtain the combined effect of the ingredients; (5) to add to the ease of administration; (6) to form new compounds.

The first, second, fourth, and fifth conditions explain themselves.

To illustrate the third: Suppose a patient suffers at the same time with malaria and a cough or cold. By combining, say, quinine and a suitable expectorant, we may control both conditions with one remedy.

To illustrate the sixth: We use daily the combination of potassium iodid and bichlorid of mercury, this giving us the red iodid of mercury, etc.

Under this head, however, we must be careful to employ no remedies which are incompatible with one another.

By **incompatibles** we may mean any of three classes of compounds, viz.:

1. Unsightly or poisonous mixtures (PHARMACEUTIC INCOMPATIBILITY).

ESSENTIALS OF PRESCRIPTION WRITING. 79

2. The formation of new compounds, or the decomposition of the ingredients, which may cause the remedy to become inert, poisonous, or explosive (CHEMIC INCOMPATIBILITY).

3. The combination of remedies having opposite therapeutic effects (PHYSIOLOGIC INCOMPATIBILITY).

Incompatibles of the first class include combinations of oils, balsams, resins (or their alcoholic solutions), tinctures (alcoholic or hydro-alcoholic), fluid extracts (alcoholic), with aqueous solutions, or *vice versâ*.

The second class includes combinations of alkalies (hydrates and carbonates), with acids, oxidizers (peroxids and peracids), with tannin, sugar, sulphur, starch, and sulphids, glycerin, alcohols, and ether; as, for instance, potassium chlorate, or potassium permanganate, with any of the above substances.

Metallic salts with acids, causing precipitation.

Mineral acids with carbonates, acetates, citrates, and salts of the vegetable acids generally.

All alkaloids and their salts are incompatible with tannic acid and all preparations containing it.

Alkaloid salts are incompatible with the alkalies and many of their salts.

Fixed oils and oleoresins may only be employed with water in the form of an emulsion.

Essential oils are soluble in water only to the extent of ♏j to f℥j.

Physiologic incompatibility would include those

FOR THERAPEUTIC NOTES.

compounds containing drugs having opposite actions ; as, for instance, a large amount of caffein to keep the patient awake and a correspondingly large amount of opium to put him to sleep.

The following combinations, which might be desirable from a therapeutic standpoint, are incompatible:

Ammonium carbonate, with *syrup of squill,* (which contains *acetic acid*). The *salts* of the heavy metals, *iron, mercury, magnesium,* etc., and the *alkaline earths, lime,* etc., are incompatible with *arsenic.* For example: **Lime-water,** with the *tincture of chlorid of iron,* solutions of *mercurial salts,* etc. **Liquor potassium arsenitis,** with *lime-water,* etc.

Belladonna and Atropin.—*Alkalies* precipitate atropin from solutions of belladonna. *Tannic acid* forms an insoluble tannate with atropin.

Iodin and the **iodids** are incompatible with *acids* and their salts, and form a heavy insoluble iodid with most **alkaloids,** which settles to the bottom of the bottle, and so there may be taken in one dose the amount intended for a number of doses. Also incompatible with *soluble metallic salts.*

Iron, or its preparations, form insoluble precipitates with **tannic acid** in any form. Hence we can use only the simple bitters and never any of the vegetable preparations containing tannic acid in combination therewith.

Mercury.—Calomel is converted into **corrosive sublimate** when combined with hydrochloric or nitro-hydrochloric acid, the chlorids, and most of the coal-tars (phenacetin, antipyrin, etc.). *Corrosive sublimate* is extremely easily decomposed, and *iodin* changes **green mercurous iodid** into the more active **red iodid.**

Nux Vomica.—Strychnine in solution forms with potassium iodid a precipitate which carries all the strychnia to the bottom of the bottle, rendering one liable to take the whole amount in one dose. (Death has been caused by this combination.)

Opium, Morphine.—Alkalies precipitate morphin from solutions of opium. **Tannic acid** forms the very slowly soluble tannate of morphin.

Pepsin, Ingluvin, etc.—**Alcohol** destroys their active properties, as do **alkalies. Acids** aid their action if diluted.

Quinine, cinchona.—Alkalies precipitate alkaloids of cinchona. *Tannic, gallic, and tartaric acids* form insoluble compounds with them.

Tannic acid and preparations containing it are incompatible with all *alkaloids* and preparations containing alkaloids; with *iron* and preparations containing it.

PART II.

CHAPTER I.

MATERIA MEDICA.

Pharmacology is sometimes defined to mean the action of drugs on the tissues of the living organism; *i. e.*, **physiologic action.** It is also sometimes supposed to include materia medica, pharmacy, and therapeutics.

Materia medica treats of all the substances, either natural or artificial, used in the practice of medicine. The University of Pennsylvania requires:

1. English name.
2. Scientific name.
3. Physical characteristics.
4. Chemic constituents.
5. Incompatibles.
6. Antidotes.
7. Preparations, United States Pharmacopeia.
8. Doses (crude drug and its preparations).
9. Adulterations, etc.
10. Habitat.

Pharmacy is the science of preparing, compounding, and dispensing medicines.

Therapeutics is the study of the action of medicines in health and disease.

A **pharmacopeia** contains explicit directions for the preparation of medicines, insuring uniformity of strength, action, and nomenclature. It may be officinal (United States) or official. An official pharmacopeia is one sanctioned by the Government; whereas, in the United States it is not sanctioned by the Government but is revised every ten years by representatives of the various recognized schools of medicine, colleges of pharmacy, of the Army and Navy Medical Corps, and of the Marine Hospital Service. Such a pharmacopeia and its preparations are called officinal.

The pharmacopeia does not recognize drugs published under a private formula or whose manufacture is restricted to one or more firms.

The **dispensatory** is much more explicit, giving the physical and chemic history, mode of preparation, doses, and therapeutics of the various drugs, and includes all remedies of whatever nature intended to cure disease.

The principal dispensatories are the "United States Dispensatory" and the "National Dispensatory."

ESSENTIALS OF MATERIA MEDICA. 89

OFFICINAL PREPARATIONS.

Decocta (decoctions) are made by *boiling* the crude drug in water. Useless if the active principle is decomposed by heat or is volatile. If starch is present in any amount it would decompose and thus be useless.

The United States Pharmacopeia recognizes two:

Decoctum—
    Cetrariæ.

Decoctum—
    Sarsaparillæ compositum.

Infusa (infusions) are made by *macerating*, percolating, or displacing the drug in *hot* or *cold water without* boiling. Officinal, four:

Infusum—
    Cinchonæ.
    Digitalis.

Infusum—
    Pruni virginianæ.
    Sennæ compositum.

Liquores (solutions) are solutions of active *non-volatile* principles in *water*. Officinal, 24:

Liquor—
- Acidi arsenosi.
- Ammonii acetatis.
- Arseni et hydrargyri iodidi.
- Calcis.
- Ferri acetatis.
- Ferri chloridi.
- Ferri citratis.
- Ferri et ammonii acetatis.
- Ferri nitratis.
- Ferri subsulphatis.
- Ferri tersulphatis.
- Hydrargyri nitratis.

Liquor—
- Iodi compositus.
- Magnesii citratis.
- Plumbi subacetatis.
- Plumbi subacetatis dilutus.
- Potassæ.
- Potassii arsenitis.
- Potassii citratis.
- Sodæ.
- Sodæ chloratæ.
- Sodii arsenatis.
- Sodii silicatis.
- Zinci chloridi.

7

ESSENTIALS OF MATERIA MEDICA. 91

Aquæ (waters) are solutions of *volatile* principles in water. Officinal, 18:

Aqua—
Ammoniæ.
Ammoniæ fortior.
Amygdalæ amaræ.
Anisi.
Aurantii florum.
Aurantii florum fortior.
Camphoræ.
Chlori.
Chloroformi.

Aqua—
Cinnamomi.
Creosoti.
Destillata.
Fœniculi.
Hydrogenii dioxidi.
Menthæ piperitæ.
Menthæ viridis.
Rosæ.
Rosæ fortior.

Spiritus (spirits) are *alcoholic* solutions of *volatile* principles. Officinal, 25:

Spiritus—
Ætheris.
Ætheris compositus.
Ætheris nitrosi.
Ammoniæ.
Ammoniæ aromaticus.
Amygdalæ amaræ.
Anisi.
Aurantii.
Aurantii compositus.
Camphoræ.
Chloroformi.
Cinnamomi.
Frumenti.

Spiritus—
Gaultheriæ.
Glonoini.
Juniperi.
Juniperi compositus.
Lavandulæ.
Limonis.
Menthæ piperitæ.
Menthæ viridis.
Myrciæ.
Myristicæ.
Phosphori.
Vini Gallici.

Tinctura (tinctures) are *alcoholic* solutions of *non-volatile* principles. Officinal, 72:

Tinctura—	Per Cent.	Tinctura—	Per Cent.
Aconiti,	35	Aloes et myrrhæ,	10
Aloes,	10	Arnicæ florum,	20

ESSENTIALS OF MATERIA MEDICA. 93

Tinctura—	Per Cent.	Tinctura—	Per Cent.
Arnicæ radicis,	10	Gelsemii,	15
Asafœtidæ,	20	Gentianæ composita,	10
Aurantii amari,	20	Guaiaci,	20
Aurantii dulcis,	20	Guaiaci ammoniata,	20
Belladonnæ (Pharm., 1880),	15	Humuli,	20
		Hydrastis,	20
Belladonnæ foliorum,	15	Hyoscyami,	15
Benzoini,	20	Iodi,	7
Benzoini composita,	12	Ipecacuanhæ et opii,	100
Bryoniæ,	10	Kino,	10
Calendulæ,	20	Krameriæ,	20
Calumbæ,	10	Lactucarii,	50
Cannabis indicæ,	15	Lavandulæ composita,	8
Cantharidis,	5	Lobeliæ,	20
Capsici,	5	Matico,	10
Cardamomi,	10	Moschi,	5
Cardamomi composita,	2	Myrrhæ,	20
Catechu composita,	10	Nucis vomicæ,	2
Chiratæ,	10	Opii,	10
Cimicifugæ,	20	Opii camphorata, less than	½
Cinchonæ,	20		
Cinchonæ composita,	10	Opii deodorati,	10
Cinnamomi,	10	Physostigmatis,	15
Colchici (Pharm., 1880).	15	Pyrethri,	20
Colchici seminis,	15	Quassiæ,	10
Croci,	10	Quillajæ,	20
Cubebæ,	20	Rhei,	10
Digitalis,	15	Rhei aromatica,	20
Tincturæ herbarum recentium,	50	Rhei dulcis,	10
		Sanguinariæ,	15
Tinctura—		Saponis viridis (Pharm, 1880),	
Ferri chloridi,	25		
Gallæ,	20	Scillæ,	15

Tinctura—	Per Cent.	Tinctura—	Per Cent.
Serpentariæ,	10	Valerianæ,	20
Stramonii(Pharm.,1880),	15	Valerianæ ammoniata,	20
Stramonii seminis,	15	Vanillæ,	10
Strophanthi,	5	Veratri viridis,	40
Sumbul,	10	Zingiberis,	20
Tolutana,	10		

The misturæ (mixtures) are simply preparations containing an insoluble substance which is held in suspension in water, usually by the aid of some viscid material. These form in some cases practically an emulsion, and some of them are so called by the 1890 United States Pharmacopeia; whereas in 1880 they were called mixtures; viz., the 1890 Pharmacopeia recognizes five mixtures:

Mistura—
  Cretæ.
  Ferri composita.
  Glycyrrhizæ composita.

Mistura—
  Potassii citratis.
  Rhei et sodæ.

In the 1880 Pharmacopeia the following were called mixtures, but the 1890 Pharmacopeia defines them as EMULSIONS. Officinal, four:

Emulsum—
  Ammoniaci.
  Amygdalæ.

Emulsum—
  Asafœtidæ.
  Chloroformi.

Mucilagines (mucilages) are watery solutions of gummy substances. There are four officinal mucilages:

Mucilago—
  Acaciæ.
  Sassafras medullæ.

Mucilago—
  Tragacanthæ.
  Ulmi.

Syrupi (syrups) are mostly watery solutions of sugary substances, though a few contain dilute acetic acid. Officinal, 31 :

Syrupus—
  Acaciæ.
  Acidi citrici.
  Acidi hydriodici.
  Allii.
  Althææ.
  Amygdalæ.
  Aurantii.
  Aurantii florum.
  Calcii lactophosphatis.
  Calcis.
  Ferri iodidi.
  Ferri, quininæ, et strychninæ phosphatum.
  Hypophosphitum.
  Hypophosphitum cum ferro.
  Ipecacuanhæ.

Syrupus—
  Krameriæ.
  Lactucarii.
  Picis liquidæ.
  Pruni virginianæ.
  Rhei.
  Rhei aromaticus.
  Rosæ.
  Rubi.
  Rubi idæi.
  Sarsaparillæ compositus.
  Scillæ.
  Scillæ compositus.
  Senegæ.
  Sennæ.
  Tolutanus.
  Zingiberis.

Mellita (honeys) have for their basis the honey of the ordinary honey-bee (*Apis mellifica*). Officinal, two :

Mel despumatum.
  Rosæ.

Aceta (vinegars) are preparations whose menstruum is dilute acetic acid or vinegar. Officinal, two :

Acetum opii.
  Scillæ.

Vina (wines), menstruum of which is white wine, containing 20 to 25 per cent. alcohol. Officinal, ten :

Vinum—
  Album.
  Antimonii.
  Colchici radicis.
  Colchici seminis.
  Ergotæ.

Vinum—
  Ferri amarum.
  Ferri citratis.
  Ipecacuanhæ.
  Opii.
  Rubrum.

Glycerita (glycerites) are solutions in glycerin. Officinal, six :

Glyceritum—
  Acidi carbolici.
  Acidi tannici.
  Amyli.

Glyceritum—
  Boroglycerini.
  Hydrastis.
  Vitelli.

Olei (oils) are volatile or non-volatile, obtained, as a rule, by distillation of plants, although a few are obtained by expression. Officinal, 50 :

Oleum—
  Adipis.
  Æthereum.
  Amygdalæ amaræ.
  Amygdalæ expressum.
  Anisi.
  Aurantii corticis.
  Aurantii florum.
  Bergamii (Pharm., 1880).
  Bergamottæ.
  Betulæ volatile.
  Cadinum (*Juniperi empy-*
  Cajuputi.   [*reumaticum*).
  Cari.

Oleum—
  Caryophylli.
  Chenopodii.
  Cinnamomi.
  Copaibæ.
  Coriandri.
  Cubebæ.
  Erigerontis.
  Eucalypti.
  Fœniculi.
  Gaultheriæ.
  Gossypii seminis.
  Hedeomæ.
  Jecoris aselli (*Morrhuæ*).

FOR THERAPEUTIC NOTES.

Oleum—
　Juniperi.
　Juniperi empyreumaticum
　　(*Cadinum*).
　Lavandulæ florum.
　Limonis.
　Lini.
　Menthæ piperitæ.
　Menthæ viridis.
　Morrhuæ (*Jecoris aselli*).
　Myrciæ.
　Myristicæ.
　Olivæ.
　Phosphoratum.
　Picis liquidæ.
　Pimentæ.

Oleum—
　Ricini.
　Rosæ.
　Rosmarini.
　Sabinæ.
　Santali.
　Sassafras.
　Sesami.
　Sinapis volatile.
　Terebinthinæ.
　Terebinthinæ rectificatum.
　Theobromatis (Theobromæ, Pharm., 1880).
　Thymi.
　Tiglii.

Test for Volatile or Non-volatile Oils.—The volatile oils evaporate entirely if a drop is placed on paper, leaving no greasy mark as do the fixed, non-volatile oils.

Oleoresinæ (oleoresins) are ethereal extracts of drugs containing an oil and a resin. Officinal, six:

Oleoresina—
　Aspidii.
　Capsici.
　Cubebæ.

Oleoresina—
　Lupulini.
　Piperis.
　Zingiberis.

Elixirs (indeclinable) are usually made with a menstruum of dilute alcohol, as are also the **succi**, or fruit juices (which are juices of the fresh fruits with enough alcohol to preserve them), as succus limonis. Two elixirs are officinal:

Elixir aromaticum.
　Phosphori.

Resinæ (resins) are made by adding water to the saturated alcoholic solutions of the resins, thus causing a precipitation. They are soluble in alcohol but not in water. Officinal, five:

Resina—
   (Residue of turpentine.)
   Copaibæ.
   Jalapæ.

Resina—
   Podophylli.
   Scammonii.

Confectiones (confections) are remedies incorporated into the form of candy. Officinal, two:

Confectio rosæ.
   Sennæ.

Trochisci (troches, lozenges, pellets) are small masses of various shapes made to dissolve slowly in the mouth, mainly to medicate the throat and mouth. Officinal, 15:

Trochisci—
   Acidi tannici.
   Ammonii chloridi.
   Catechu.
   Cretæ.
   Cubebæ.
   Ferri.
   Glycyrrhizæ et opii.
   Ipecacuanhæ.

Trochisci—
   Krameriæ.
   Menthæ piperitæ.
   Morphinæ et ipecacuanhæ.
   Potassii chloratis.
   Santonini.
   Sodii bicarbonatis.
   Zingiberis.

Suppositoria (suppositories) are conic masses of oil of theobroma (butter of cacao) combined with various medicaments, intended for use in the rectum, vagina, and urethra. Rectal and urethral supposi-

tories should weigh about one gram ; for vaginal use, three grams. Officinal :
Suppositoria glycerini.

Unguenta (ointments) contain of yellow wax 20 per cent., of lard 80 per cent. ; for external use only. Officinal, 22 :

Unguentum—
   Acidi carbolici.
   Acidi tannici.
   Aquæ rosæ.
   Belladonnæ.
   Chrysarobini.
   Diachylon.
   Gallæ.
   Hydrargyri.
   Hydrargyri ammoniati.
   Hydrargyri nitratis.
   Hydrargyri oxidi flavi.

Unguentum—
   Hydrargyri oxidi rubri.
   Iodi.
   Iodoformi.
   Picis liquidæ.
   Plumbi carbonatis.
   Plumbi iodidi.
   Potassii iodidi.
   Stramonii.
   Sulphuris.
   Veratrinæ.
   Zinci oxidi.

Cerata (cerates) contain of white wax 30 per cent. and of lard 70 per cent. Officinal, five :

Ceratum—
   Camphoræ.
   Cantharidis.
   Cetacei.

Ceratum—
   Plumbi subacetatis.
   Resinæ.

Extracta (extracts) are made by evaporating solutions of vegetable substances, sometimes the fresh juice, resulting in a soft mass. Officinal, 33 :

Extractum—
   Aconiti.
   Aloes.
   Arnicæ radicis.

Extractum—
   Belladonnæ foliorum alco-
     holicum.
   Cannabis indicæ.

Extractum—
  Cimicifugæ.
  Cinchonæ.
  Colchici radicis.
  Colocynthidis.
  Colocynthidis compositum.
  Conii.
  Digitalis.
  Ergotæ.
  Euonymi.
  Gentianæ.
  Glycyrrhizæ.
  Glycyrrhizæ purum.
  Hæmatoxyli.
  Hyoscyami.

Extractum—
  Iridis.
  Jalapæ.
  Juglandis.
  Krameriæ.
  Leptandræ.
  Nucis vomicæ.
  Opii.
  Physostigmatis.
  Podophylli.
  Quassiæ.
  Rhei.
  Stramonii seminis.
  Taraxaci.
  Uvæ ursi.

Extracta fluida (fluid extracts) are, as a rule, the most powerful of the liquid preparations, and generally *one minim is equal to one grain of the crude drug*. Officinal, 88 :

Extractum—
  Aconiti fluidum.
  Apocyni fluidum.
  Arnicæ radicis fluidum.
  Aromaticum fluidum.
  Asclepiadis fluidum.
  Aspidospermatis fluidum.
  Aurantii amari fluidum.
  Belladonnæ radicis fluidum
    (Belladonnæ fluidum, Pharm., 1880.)
  Buchu fluidum.
  Calami fluidum.
  Calumbæ fluidum.

Extractum—
  Cannabis indicæ fluidum.
  Capsici fluidum.
  Castaneæ fluidum.
  Chimaphilæ fluidum.
  Chiratæ fluidum.
  Cimicifugæ fluidum.
  Cinchonæ fluidum.
  Cocæ fluidum.
  Colchici radicis fluidum.
  Colchici seminis fluidum.
  Conii fluidum.
  Convallariæ fluidum.
  Cubebæ fluidum.

ESSENTIALS OF MATERIA MEDICA. 109

Extractum—
Cusso fluidum.
Cypripedii fluidum.
Digitalis fluidum.
Dulcamaræ fluidum.
Ergotæ fluidum.
Eriodictyi fluidum.
Erythroxyli fluidum
  (Pharm., 1880).
Eucalypti fluidum.
Eupatorii fluidum.
Frangulæ fluidum.
Gelsemii fluidum.
Gentianæ fluidum.
Geranii fluidum.
Glycyrrhizæ fluidum.
Gossypii radicis fluidum.
Grindeliæ fluidum.
Guaranæ fluidum.
Hamamelidis fluidum
Hydrastis fluidum.
Hyoscyami fluidum.
Ipecacuanhæ fluidum.
Iridis fluidum.
Krameriæ fluidum.
Lappæ fluidum.
Leptandræ fluidum.
Lobeliæ fluidum.
Lupulini fluidum.
Matico fluidum.
Menispermi fluidum.
Mezerei fluidum.
Nucis vomicæ fluidum.
Pareiræ fluidum.

Extractum—
Phytolaccæ radicis fluidum.
Pilocarpi fluidum.
Podophylli fluidum.
Pruni virginianæ fluidum.
Quassiæ fluidum.
Rhamni purshianæ fluidum.
Rhei fluidum.
Rhois glabræ fluidum.
Rosæ fluidum.
Rubi fluidum.
Rumicis fluidum.
Sabinæ fluidum.
Sanguinariæ fluidum.
Sarsaparillæ fluidum.
Sarsaparillæ fluidum compositum.
Scillæ fluidum.
Scoparii fluidum.
Scutellariæ fluidum.
Senegæ fluidum.
Sennæ fluidum.
Serpentariæ fluidum.
Spigeliæ fluidum.
Stillingiæ fluidum.
Stramonii seminis fluidum (Stramonii fluidum, Pharm., 1880.)
Taraxaci fluidum.
Tritici fluidum.
Uvæ ursi fluidum.
Valerianæ fluidum.
Veratri viridis fluidum.

FOR THERAPEUTIC NOTES.

Extractum—
　Viburni opuli fluidum.
　Viburni prunifolii fluidum.

Extractum—
　Xanthoxyli fluidum.
　Zingiberis fluidum.

Emplastra (plasters) are prepared by spreading—by means of heat—on muslin, silk, etc., the material intended for external application. Officinal, 13:

Emplastrum—
　Ammoniaci cum hydrargyro.
　Arnicæ.
　Belladonnæ.
　Capsici.
　Ferri.
　Hydrargyri.

Emplastrum—
　Ichthyocollæ.
　Opii.
　Picis burgundicæ.
　Picis cantharidatum.
　Plumbi.
　Resinæ.
　Saponis.

Pilulæ (pills) are small rounded masses to be swallowed whole. They include only such substances as are (1) of bad taste; (2) of small dose; (3) of drugs having slow action; (4) of substances too heavy for suspension in liquids, or insoluble therein. For obvious reasons we can not use in pill form substances—(1) of large dose; (2) when rapid action is desired (as emetics, etc.); (3) corrosive substances; (4) deliquescent salts. Officinal, 15:

Pilulæ—
　Aloes.
　Aloes et asafœtidæ.
　Aloes et ferri.
　Aloes et mastiches.
　Aloes et myrrhæ.
　Antimonii compositæ.
　Asafœtidæ.
　Catharticæ compositæ.

Pilulæ—
　Catharticæ vegetabiles.
　Ferri carbonatis.
　Ferri iodidi.
　Opii.
　Phosphori.
　Rhei.
　Rhei compositæ.

Pulveres (powders) are remedies, always dry substances, dispensed in small papers, each of which contains one dose. It must be borne in mind that deliquescent salts, volatile substances, and substances which liquefy when brought together,—chloral and camphor, for example,—are unfit for use in this manner. Officinal, nine:

Pulvis—
 Antimonialis.
 Aromaticus.
 Cretæ compositus.
 Effervescens compositus.
 Glycyrrhizæ compositus.

Pulvis—
 Ipecacuanhæ et opii.
 Jalapæ compositus (*Purgans*).
 Morphinæ compositus.
 Rhei compositus.

Chartæ (papers) are medicated papers, as a rule, to be burned in order that the vapor of drug they contain may be inhaled, or for external application. Officinal, two:

Charta potassii nitratis.
 Sinapis.

Linimenta (liniments) are intended to be used externally, as a rule, to be applied with friction, and are generally of a soapy or oily consistency. Officinal, nine:

Linimentum—
 Ammoniæ.
 Belladonnæ.
 Calcis.
 Camphoræ.
 Chloroformi.

Linimentum—
 Saponis.
 Saponis mollis.
 Sinapis compositum.
 Terebinthinæ.

Collodium (collodions).—Solutions of ether in gun-cotton which rapidly evaporate when applied to the skin, forming a translucent film containing the medicament and protecting the part to which it is applied. Officinal, three:

Collodium cantharidatum.
  Flexile.
  Stypticum.

Beside the above-mentioned officinal preparations we have the following

NON-OFFICINAL PREPARATIONS.

Enema (enema or clyster).—Liquids to be injected per rectum.

Bougia.—Small cylinders of cacao butter mixed with the remedy to be used in uterus and urethra.

Pesoaria.—Vaginal suppositories.

Granulum (granules).—Very small pills containing powerful drugs.

Dragées.—Sugar-coated pills of French origin.

Cachets.—These consist of two depressed discs of flour-paper in the interior of which the remedy—generally given in large dose—is placed. These become very slippery when placed in the mouth, and large doses of drugs disagreeable to the taste may thus be taken without awakening the repugnance of the patient.

Cachous.—Small, highly perfumed pills, often covered with gold- or silver-foil; used mostly to perfume the breath.

CHAPTER II.

AVERAGE DOSES.

Although, of course, we can not give one definite, absolute dose for any drug or drugs, still, for purposes of study, the following list will be found sufficiently accurate. Naturally, the student understands that the dose must be modified to suit the case in hand, as explained in a previous chapter.

Crude drugs may be given in gr. v–x doses.

Exceptions, poisons (see list under Tinctures and Fluid extracts); dose when used, gr. j, about.

Extracts may be given in doses of gr. j–iij.

Exceptions, poisons (see list under Tinctures and Fluid extracts); dose, gr. ⅛–½.

Fluid extracts may be given in ♏x doses.

Exceptions, POISONS: fluid extracts of *aconite, belladonna, digitalis, squill, stramonium, veratrum viride,* the dose of all these being ♏j–ij; and the fluid extracts of *colchicum seed,* ♏ij–x; *colchicum root,* ♏ij–v; *sanguinaria,* ♏j–v; *nux vomica,* ♏j–v.

Tinctures may be given in doses of fʒss–ij.

Exceptions, POISONS: tinctures of *digitalis, iodine, nux vomica, opium,* and *deodorized tincture of opium,* ♏ij–x; *aconite,* ♏j–iij; *belladonna,* ♏x–xx; *col-*

chicum, ♏v–xxx ; *physostigma*, ♏x–xv ; *squill*, ♏v–xxx.

Infusions and decoctions may be given in f℥ss–ij doses, except infusion of *digitalis*, f℈j–iv.

Syrups may be given in f℈j–ij doses, except the syrups of *iron* and *compound syrup of squill*, ♏v–xxx.

Mixtures and emulsions may be given in f℥ss doses.

Oils may be given internally in ♏ij–v–x doses, except POISONS (*croton oil* and *phosphorated oil*, ♏j) and *castor* oil and *cod-liver* oil, f℥ss.

Alkaloids may be given in doses of gr. $\frac{1}{80}$–$\frac{1}{30}$–$\frac{1}{10}$, except *aconitine*, gr. $\frac{1}{500}$; *caffeine*, gr. ij–xv ; *quinine and its associated alkaloids*, gr. ij–xx ; *morphine and its associated alkaloids*, gr. ⅛–¼–½ ; *pelletierine and isopelletierine*, gr. v–x.

NOTE.—An alkaloid (vegetable alkali) is "a vegetable extract capable of uniting with acids to form salts," as does ammonia, etc.

A glucoside is "an organic substance which is resolvable by the presence of acids, or other slight chemical influence, into glucose and some other proximate principle" (Example: Certain varieties of tannic acid form glucose and gallic acid, etc.).

CHAPTER III.

OFFICINAL DRUGS AND PREPARATIONS.— IMPORTANT NON-OFFICINAL PREPARATIONS.—DOSES.

The study of materia medica renders necessary the knowledge of the following points in reference to each and every substance employed therein, viz. :
1. English name.
2. Scientific name.
3. Physical characteristics.
4. Chemic constituents.
5. Incompatibles (if any).
6. Antidotes (if poisonous).
7. Preparations (United States Pharmacopeia).
8. Doses (of crude drug and its preparations).
9. Adulterations.
10. Habitat.
11. How made (of chemicals).
12. Parts of plant used (if of vegetable kingdom).
13. Parts of animal used (if animal).

The classification adopted by Dr. H. C. Wood in "Therapeutics : Its Principles and Practice," will be followed in the consideration of the drugs to follow:

Division I.—**Systemic Remedies.**
Division II —**Extraneous Remedies.**

SYSTEMIC REMEDIES

May again be divided into—

(*a*) *General remedies*, "affecting the tissues of the body generally, or such organized systems as reach all portions of the body."

(*b*) *Local remedies*, those affecting only one organ of the body.

The **general remedies** may be divided into the following three orders:

I. Nervines, acting on the nervous system.
II. Cardiants, acting on the circulation.
III. Nutrients, acting on the general nutrition.

NERVINES.

A. Medicines acting on the cerebrum.
B. Medicines acting on the remainder of the nervous system.

A. FAMILY 1.—**Antispasmodics**, employed for the relief of minor spasms and nervous manifestations, are feeble cerebral stimulants.

FAMILY 2.—**Anesthetics** are used to produce anesthesia (local and general).

FAMILY 3 —**Somnifacients**, in proper doses produce sleep without delirium.

FAMILY 4 —**Delirifacients**, in proper doses produce delirium first, then stupor.

B. FAMILY 5.—**Excito-motors**—drugs causing violent tetanic spasms in overdose.

FOR THERAPEUTIC NOTES.

FAMILY 6.—**Depresso-motors**—drugs causing paralysis in overdose.

CARDIANTS.

FAMILY 1.—**Cardiac stimulants**—drugs which increase arterial pressure.

FAMILY 2.—**Cardiac depressants**—drugs which decrease arterial pressure.

NUTRIENTS.

FAMILY 1.—**Astringents**—drugs causing contraction of various organs.

FAMILY 2.—**Tonics**—drugs which increase nutrition and vital power.

FAMILY 3.—**Alteratives**—drugs which modify nutrition and overcome "certain chronic pathologic processes."

FAMILY 4.—**Antiperiodics**—drugs which overcome the effects of malarial poisoning.

FAMILY 5.—**Antipyretics**—drugs which overcome febrile movements.

Local Remedies.
FAMILY 1.—Stomachics.
FAMILY 2 —Emetics.
FAMILY 3.—Cathartics.
FAMILY 4.—Diuretics.
FAMILY 5.—Diaphoretics.
FAMILY 6.—Expectorants.
FAMILY 7.—Emmenagogues.
FAMILY 8.—Oxytocics.
FAMILY 9.—Sialagogues.
FAMILY 10.—Errhines.
FAMILY 11.—Epispastics.
FAMILY 12.—Rubefacients.
FAMILY 13.—Escharotics.
FAMILY 14 —Demulcents.
FAMILY 15.—Emollients.
FAMILY 16.—Protectives.

EXTRANEOUS REMEDIES.
FAMILY 1.—Antacids.
FAMILY 2.—Anthelmintics.
FAMILY 3.—Digestants.
FAMILY 4.—Absorbents.
FAMILY 5.—Disinfectants.

FOR THERAPEUTIC NOTES.

ORDER I.—NERVINES.

FAMILY I.—ANTISPASMODICS.

Drugs capable of controlling minor spasms of voluntary or involuntary muscles.

Officinal Name, MOSCHUS. *English Name*, MUSK.

Definition.—Dried *secretions* from the *preputial follicles* of Moschus moschiferus (musk deer). *Class.*—Mammalia. *Order.*—Ruminantia. *Habitat.*—Thibet. *Physical Properties.*—Irregular, crumbly grains; dark red-brown; peculiar, penetrating, persistent odor; bitter taste.

The crude drug is given in **doses** of gr. v–xv in emulsion, capsule, or injection (per rectum).

Officinal Preparation.

Tinctura Moschi, ... f ℨ ss–ij.

Officinal Name, VALERIANA. *English Name*, VALERIAN.

Definition.—The *rhizome* and *rootlets* of Valeriana officinalis.

Natural Order.—Valerianeæ. *Habitat.*—Europe. *Physical Properties.*—Its peculiar odor, resembling perspiration (due to *valerianic acid*), becomes stronger on keeping; taste camphoraceous and bitter. Contains *valerianic acid* and *oil of valerian*, both volatile.

Valerian root
root or good powdered

Valerianic Acid =
$H \cdot C_5 H_9 O_2$

$N H_4 \cdot C_5 H_9 O_2 =$
Ammon. Valerian
5 to 10 up to 15

Officinal Preparations.
Extractum Valerianæ Fluidum, . . f ℥ ss–j.
Tinctura Valerianæ, ... f ℥ ss–ij.
Tinctura Valerianæ Ammoniata, . . f ℥ ss–ij.
Ammonii Valerianas, ... gr. ij–x.

Iron, quinine, and *zinc valerianate* will be considered under iron, quinine, and zinc.

Officinal Name, ASAFŒTIDA. *English Name,* ASAFETIDA.

Definition.—A *gum-resin* obtained from the *root* of Ferula fœtida.
Natural Order.—Umbelliferæ. *Habitat.*—Afghanistan. *Physical Properties.*—Irregular masses of whitish tears imbedded in a yellowish-gray, sticky mass; has a persistent, alliaceous odor, and bitter, acrid, alliaceous taste. Contains a volatile oil, a gum, and a resin.

*Officinal Preparations.**
Emulsum or Mistura (Pharm., 1880)
  Asafœtidæ, ... f ℥ ss–j.
Pilulæ Aloes et Asafœtidæ (aloes, asafetida, and soap, of each gr. 1⅓).
Pilulæ Asafœtidæ, each, ... gr. iij.
Tinctura Asafœtidæ, ..., . . f ℥ ss–ij.

Officinal Name, CAMPHORA. *English Name,* CAMPHOR.

Definition.—A stearopten (having the nature of a ketone) from the Cinnamomum camphora.

---

* Mistura magnesiæ et asafœtidæ (Dewees' Carminative), no longer officinal, contain mag. carbonate, seven per cent.; laudanum, one per cent.; tinct. asafœtidæ, seven per cent.; sugar and water.

$C_{10}H_{16}O$ Camphor *ail*

7. *bark* in water —

should not be call
gum camph. Gum,
stick, mucilage—
Character
imperceptible H_2O.

Natural Order.—Laurineæ. *Habitat.*—China and Japan. *Physical Properties.*—White, tough, translucent masses; crystalline, penetrating, characteristic odor, and pungent, aromatic taste. Pulverizable in presence of alcohol, ether, or chloroform.

Officinal Preparations.

Aqua Camphoræ (1 to 125), ... f℥ ss–ij.
Linimentum Camphoræ (20 per cent. solution in cotton-seed oil).
Linimentum Saponis.
Spiritus Camphoræ (ten per cent. camphor), ... gtt. v–f℥j.
Tinctura Opii Camphorata (see Opium).
Camphora Monobromata, ... gr. iij–xv.

Officinal Name, SPIRITUS ÆTHERIS COMPOSITUS.
English Name, COMPOUND SPIRIT OF ETHER
(HOFFMANN'S ANODYNE).

Definition.—An alcoholic solution of ether and ethereal oil made as follows:

Ether, ... 325 c. c.
Alcohol, ... 650 "
Ethereal Oil, ... 25 "
                                    _____
                                    1000 c. c.

The genuine preparation imparts a cloudiness to water when about 45 drops of it are added to a pint of water. Adulterations remain clear, owing to absence of oil.

Dose, f℥ss–ij.

FOR THERAPEUTIC NOTES.

Officinal Name, HUMULUS. *English Name*, HOPS.

Definition.—*Strobiles* of Humulus lupulus. *Natural Order.*—Urticaceæ. *Habitat.*—Europe and North America. *Physical Properties.*—Ovate scales containing a glandular, yellowish powder called *lupulin*, to which its activity is due. Scales themselves are greenish, have an aromatic odor, bitter, astringent taste. Crude drug used only for poultices.

Officinal Preparations.

Tinctura Humuli,	f ℥ ss–ij.
Lupulinum,	gr. x–xx.
Extractum Lupulini Fluidum,	f ℥ ss–j.
Oleoresina Lupulini,	♏ v–xxx.

Officinal Name, LACTUCARIUM. *English Name*, LACTUCARIUM, LETTUCE OPIUM.

Definition.—Concrete *milky juice* of Lactuca virosa.

Natural Order.— Compositæ. *Habitat.*—Indigenous.

Dose of crude drug, gr. x–ʒj.

Officinal Preparation.

Tinctura Lactucarii, ... f ℥ ss–f ℥ ij.

Officinal Name, CIMICIFUGA. *English Name*, BLACK SNAKEROOT, BLACK COHOSH.

Definition.—*Rhizome* and *rootlets* of Cimicifuga racemosa.

Natural Order. — Ranunculaceæ. *Habitat.* — United States.

Castanea — leaves
of " dentata,
ordinary chestnut
leaves, tannic acid
treatment of whoop-
ing cough — a decoction.

Extracter Fluidum
Castanea — ordinary
in use.

Officinal Preparations.
Extractum Cimicifugæ, ... gr. ¼ -iij.
Extractum Cimicifugæ Fluidum, . . . f ℥ ss–j.
Tinctura Cimicifugæ, ... f ℥ ss–ij.

FAMILY II.—ANESTHETICS.

General anesthetics produce loss of consciousness of entire body.

Local anesthetics cause loss of sensation only in the part of the body to which they are applied,— practically only the mucous surfaces and superficial regions of the body.

NITROUS OXID (N_2O), LAUGHING GAS, NITROGEN MONOXID.*

Definition.—Colorless, practically odorless, gas, made by heating ammonium nitrate.

$$NO_3NH_4 + \text{heat} = N_2O + 2H_2O.$$

Used mainly in dentistry. Administered by inhalation.

Officinal Name, ÆTHER. *English Name,* ETHER, SULPHURIC ETHER (*incorrect*), ÆTHER FORTIOR
· (Pharm., 1880).

Definition.—A volatile liquid prepared by distilling alcohol in presence of sulphuric acid. Contains 96 per cent., by weight, of ethyl oxide [$(C_2H_5)_2O$], and about four per cent. of alcohol. Specific gravity, 0.728.

---
* Not officinal.

H_5I_2 O

7th Oxide in
long ether

of ill — 2½ hour
in air.

H_5I_2, H_3O_2 — air

Physical Properties.—Transparent, colorless, mobile liquid, of characteristic odor and sweetish, burning taste. Volatile and highly inflammable. The vapor, if mixed with air and ignited, explodes violently; is heavier than air, and therefore may be used in the presence of artificial light, provided the light is kept higher than the vapor.

Officinal Preparations.

Spiritus Ætheris (30 per cent.), . . . f ℥ ss–ij.
Spiritus Ætheris Compositus, ... f ℥ ss–ij.

Officinal Name, CHLOROFORMUM (CHLOROFORMUM PURIFICATUM, Pharm., 1880). *English Name*, CHLOROFORM.

Definition.—Volatile liquid obtained by distilling alcohol in the presence of chlorinated and slaked lime. It is clear, heavy, and colorless, has a sweet taste and characteristic ethereal odor. Contains 99 per cent., by weight, of absolute chloroform and one per cent. alcohol. Specific gravity, 1.490 (about). It is not inflammable, but its vapor burns with a green flame.

Officinal Preparations.

Aqua Chloroformi, ... f ℥ ss.
Emulsum Chloroformi, ... f ℥ ss.
Linimentum Chloroformi, 40 per cent.
 chloroform in soap liniment.
Spiritus Chloroformi, ten per cent., . f ℥ ss–ij.

Bichloride of methylene, bromoform, and ethyl bromide somewhat resemble chloroform, and are occasionally used as anesthetics.

pentane C_5H_{10}

monide

illuded by the con-
sint phys.
arrow adulteration, can be
found by. Plant mineral

Local Anesthetics.—Cocaine (see Coca) is used as a local anesthetic on mucous membranes by direct application and consequent absorption, or under skin surfaces by hypodermic injection after checking the blood-supply to and from the part.

Dose of the hydrochlorate, gr. ¼.

Eucaine* is claimed to be less poisonous and dangerous than cocaine. Used in same dose and manner.

The rapid evaporation of **chloride of ethyl** produces the same effect as freezing of a part, and many small operations can be done under its influence. Contained in glass capsules, from which it is sprayed on the part until anesthesia follows.

FAMILY III.—SOMNIFACIENTS.

Somnifacients are drugs used to produce sleep.

Officinal Name, OPIUM. *English Name*, OPIUM.

Definition.—The *concrete, milky exudation* obtained by incising the unripe capsules of Papaver somniferum (poppy).

Natural Order.—Papaveraceæ. *Habitat.*—Asia Minor, Persia, India, etc. *Physical Properties.*— Irregular cakes wrapped in poppy leaves; plastic or harder; chestnut brown or darker; sharp, narcotic odor, and peculiar, bitter taste. Must yield in the

* Not officinal.

2 ½ ounces of nothing [illegible]
on Conway brides, to feed upon
bread made by Hebrew bakers.

Oil of Poppy — may be used
as a substitute of olive oil

Peculiarity of Morphine poison
is that the pupil is abnorm-
ly C[ontracted]

Mode of treatment, remove the
poison in stomach by an emetic
[illegible] 1 Tablespoon of mustard.
[illegible] Pow. Ethereal [illegible] Sulphas
Best means of keeping awake
administer Atropine
∧   Strych.——
Potassii Permanganate

normal moist condition not less than nine per cent. of **morphine**, its principal alkaloid.

Officinal.

Opii, ... Opii Pulvis (not less than 13 nor more than 15 per cent. of morphine), . Opium Deodoratum (Opium Denarcotisatum, Pharm., 1880), ...	Dose, gr. ¼–ij.

Officinal Preparations.

Acetum Opii, ... Vinum Opii, ... Tinctura Opii (laudanum), Tinctura Opii Deodorati, ... Tinctura Ipecac. et Opii (corresponds to Dover's powder), ...	ten per cent. of opium. Dose, ♏x–xx.
Tinctura Opii Camphorata (paregoric) contains two grains of opium to one fluidounce; also benzoic acid, camphor, oil of anise, glycerin, and dilute alcohol, ...	Dose, f ℥ ss–iv.

Extractum Opii, ... gr. ss.
Pilulæ Opii, each contain of powdered
  opium, ... gr. j.
Pulvis Ipecac. et Opii (*Dover's powder*)
  (O., 1 ; ip., 1 ; sacch. lactis, 8), . . gr. x.
Trochisci Opii et Glycyrrhizæ, . . . x = gr. j.
Emplastrum Opii, *ext. opii*, ... 6 in 100.

Opium contains a number of alkaloids in combination with meconic and thebolactic acids. Only **morphine** and **codeine** are officinal.

144 FOR THERAPEUTIC NOTES.

[Handwritten notes, largely illegible. Partial readings:]

...many after effects of [opium?]
...due to the Narcotine

...variety of [?] are used
...Col. of Opium

...Chloride Tincture grain
...reactions (Laudanum ?
...Meconic acid —
...[?] color.
Children [from?] Morphine may have
[?] } Chapp[?] in that
— — —   Contains no [?]
[?] + 1/3 less Morph.

.

ESSENTIALS OF MATERIA MEDICA. 145

Officinal Preparations.
Morphina, used in pharmacy.
Morphinæ Sulphas, ... ⎫
Morphinæ Acetas, ... ⎬ gr. ⅛–½.
Morphinæ Hydrochloras, ... ⎭
Pulvis Morphinæ Compositus (morph.
  sulph., 1; camphor, 19; excipient,
  80) (*Tully's powder*), ... gr. x.
Trochisci Morphinæ et Ipecacuanhæ, . 1 = gr. $\frac{1}{40}$.
Codeina, ... gr. ¼–ij.
Narceina,* Thebaina or Paramorphina,*
  Narcotina,* Laudanina,* Meconina,*
  Papaverina,* Porphyroxina,* etc., . gr. ij–x.
Apomorphinæ Hydrochloras (see Emetics).

Solutions of opium produce a blood-red color on the addition of ferric chloride (due to meconic acid). Morphine strikes a deep blue with ferric chloride and a rich orange-red, fading into yellow, with concentrated nitric acid. (See different result with quinine, under Cinchona.)

If morphine is treated with cold, concentrated sulphuric acid (free from nitric acid), on subsequent addition of a small crystal of potassium permanganate a greenish color only should be produced. Strychnine gives with this test a violet or purple color.

(For Poisoning see chap. on Antidotes.)

Officinal Name, CHLORAL. *English Name*, CHLORAL HYDRATE OR CHLORAL.

Definition.—A volatile, crystalline solid, of aro-

---

* Not officinal.

matic, acrid odor, and bitter, caustic taste. Obtained by acting on alcohol with chlorine.

Preparation.
Chloralis, ... gr. x–xxx.

Chloral forms with camphor, chloral-camphor—a liquid for external use only.

HYOSCINÆ HYDROBROMAS, gr. $\frac{1}{60}$. (See Hyoscyamus.)

METACHLORAL,* CHLORALAMID,* CHLORALOSE,* AND BUTYL CHLORAL HYDRATE,*
All modifications of chloral, are given in the same manner and dose.

SULPHONAL*
Is a valuable somnifacient in **doses** of gr. x–ʒss.

TRIONAL AND TETRONAL*
Are given in the same dose as sulphonal.

Officinal Name, PARALDEHYDUM. *English Name,* PAR-ALDEHYDE (C_2H_4O).

Definition.—Acetic aldehyde caused by oxidation, as in the effect of chromic acid on alcohol.

Dose, ♏xx–xl.

FAMILY IV.—DELIRIFACIENTS.
Remedies causing marked dilatation of the pupil, and acting on the cerebrum; causing delirium in overdose.

* Not officinal.

ESSENTIALS OF MATERIA MEDICA. 149

Officinal Name, CANNABIS INDICA. *English Name*,
INDIAN HEMP, INDIAN CANNABIS.

Definition.—The *flowering tops* of the *female* plant of the Cannabis sativa, grown in the East Indies.

Natural Order.—Urticaceæ. *Habitat.*—East Indies.

Officinal Preparations.

Extractum Cannabis Indicæ, ... gr. ¼-j.
Extractum Cannabis Indicæ Fluidum, . ♏j-x.
Tinctura Cannabis Indicæ, ... ♏x-xxx.

Officinal Name,   *English Name*,
BELLADONNÆ FOLIA,   BELLADONNA LEAVES,
BELLADONNÆ RADIX.   BELLADONNA ROOT.

Definition.—*Leaves* and *root* of Atropa belladonna (deadly nightshade), a European perennial.

Natural Order.—Solanaceæ. *Habitat.*—Europe.

Officinal Preparations.

Leaf:
   Extractum Belladonnæ Foliorum Alcoholicum, ... gr. ⅛-½.
   Tinctura Belladonnæ Foliorum, ... ♏x-xxx.

Root:
   Extractum Belladonnæ Radicis Fluidum, ♏j-ij.
   Emplastrum Belladonnæ, 2 parts of alcoholic ext. to 10 parts of plaster.
   Linimentum Belladonnæ, camphor, 50 parts, fluid extract belladonna to make 1000.

Belladonna contains the alkaloid **atropine**, officinal as **atropinæ sulphas**, gr. $\frac{1}{200}-\frac{1}{50}$.

An alcoholic solution of atropine added to mercuric chloride gives a yellow precipitate turning to red.

It may also be tested **physiologically** by applying the suspected solution to an eye of a lower animal. Atropine invariably causes dilatation of the pupil.

Atropine is incompatible with tannic acid, and alkalies precipitate atropine from the various solutions of belladonna.

Homatropine is an artificial alkaloid and is sometimes preferred as a mydriatic, its effect passing off sooner than that of atropine.

(For Poisoning see chap. on Antidotes.)

Officinal Name, STRAMONII SEMEN. *English Name*, STRAMONIUM SEED.

Definition.—The *seeds* of the Datura stramonium, or Jamestown (Jimpson) weed.

Natural Order.—Solanaceæ. *Habitat.*—United States.

Officinal Preparations.

Extractum Stramonii Seminis, . . . gr. ¼–j.
Extractum Stramonii Seminis Fluidum, ♏ j–v.
Tinctura Stramonii Seminis, ... ♏ v–xx.
Unguentum Stramonii (ten per cent. of extract).

Stramonium yields the alkaloid **daturine**, which is practically the same in action as atropine, but is not officinal.

FOR THERAPEUTIC NOTES.

Officinal Name, HYOSCYAMUS. *English Name*, HENBANE.

Definition.—The *leaves* and *flowering tops* of Hyoscyamus niger, from plants of the *second year's* growth.

Natural Order.—Solanaceæ. *Habitat.*—Europe.

Officinal Preparations.

Extractum Hyoscyami, ... gr. j–ij.
Extractum Hyoscyami Fluidum, . . . ♏v–xx.
Tinctura Hyoscyami, ... f℥ss–j.

Hyoscyamus yields the alkaloids **hyoscine** (hydrobromate), gr. $\frac{1}{150}-\frac{1}{80}$, and **hyoscyamine** (sulphate and hydrobromate), gr. $\frac{1}{80}$; both are officinal.

Officinal Name, COCA (ERYTHROXYLON, Pharm., 1880).
English Name, COCA.

Definition.—The *leaves* of the Erythroxylon coca.

Natural Order.—Lineæ. *Habitat.*—South America.

Officinal Preparation.

Extractum Cocæ Fluidum, ... f℥ss–ij.

Its alkaloid is **cocaine** (hydrochlorate), **dose**, gr. ¼–j, used mostly as a local anesthetic.

Tropacocaine, from the narrow-leaved coca of Java, resembles, for all practical purposes, cocaine in its action.

All drug have a drug
in them known as [illegible]
Strychnine

[illegible] the U.S. [illegible]
[illegible]
Bromine in Britain

FAMILY V.—EXCITO-MOTORS.

These remedies, by exciting the reflex centers of the spinal cord, produce in normal doses increased muscular activity; in poisonous doses they produce tetanic convulsions.

Officinal Name, NUX VOMICA. *English Name*, NUX VOMICA.

Definition.—The *seed* of Strychnos nux-vomica, a small East Indian tree.

Natural Order. —Loganiaceæ.   *Habitat.*— East Indies.

Officinal Preparations.

Extractum Nucis Vomicæ, ... gr. ¼–ss.
Extractum Nucis Vomicæ Fluidum, . . ℞ j–v.
Tinctura Nucis Vomicæ (20 per cent.), ℞ ij–x.

Alkaloids.—Strychnine (officinal) and **brucine** (non-officinal). Strychninæ sulphas, gr. $\frac{1}{60}-\frac{1}{20}$.

Test.—A crystal of potassium dichromate drawn through a solution of strychnine in concentrated sulphuric acid produces a blue color changing to violet, purple-red, then orange or yellow.

Brucine gives a blood-red color, fading into yellow, with nitric acid.

(For Poisoning see chap. on Antidotes.)

Officinal Name, IGNATIA (Pharm., 1880). *English Name*, IGNATIA.

Definition.—This drug, no longer officinal, is

Rel. solubility of
Stry + Brucine
67 00 Stry —
Sulphos [?]
[?] acid and an al-
- or Carbonate own
iodide of [?] —
Bitterness of Stry - a [?]
Physiological +
Chemical.

like nux vomica in action and contains also **strychnine and brucine.**

FAMILY VI.—DEPRESSO-MOTORS.

By depressing the spinal centers these remedies lessen muscular activity.

Officinal Name, PHYSOSTIGMA. *English Name,* CALABAR BEAN.

Definition.—The *seed* of Physostigma venenosum.

Natural Order.—Leguminosæ. *Habitat.*—West coast of Africa.

Contains the alkaloid **physostigmine** (or eserine).

Officinal Preparations.

Extractum Physostigmatis, ... gr. ⅛-j.
Tinctura Physostigmatis, ... ♏v-xx.

Physostigmine or Eserine, its alkaloid, is officinal as:

Physostigminæ Salicylas, . .
Physostigminæ Sulphas, ... } gr. 1/60–1/10.

The bromides belong to this class.

Potassii Bromidum, ... gr. v-ʒj.
Lithii Bromidum, ... gr. x-ʒss.
Sodii Bromidum, ... gr. v-ʒj.
Ammonii Bromidum, ... gr. v-ʒ ss.
Acidum Hydrobromicum Dilutum (ten
  per cent.), ... ʒ ss-j.

Officinal Name, AMYL NITRIS. *English Name*, AMYL NITRITE.

Definition.—A yellow, oily, very volatile liquid, containing 80 per cent. of *amyl* (mainly isoamyl). Its odor resembles that of fruit. Almost insoluble in water; miscible with alcohol or ether. Prepared by the action of nitric acid on amylic alcohol (fusel oil).

Dose, ♏j–x by inhalation; ♏j–v, internally.

Officinal Name, SPIRITUS GLONOINI. *English Name*, SPIRIT OF NITROGLYCERIN (GLONOIN).

Definition.—An alcoholic (one per cent. by weight) solution of glonoin (propenyl) trinitrate. Clear and colorless, resembling alcohol in odor and taste. Great caution is necessary in handling it, owing to its extremely explosive properties. Should the alcohol evaporate explosion may occur. Tasting even a small amount may cause violent headache. Prepared by the action of nitric acid on glycerin.

Dose, ♏ss–j.

Officinal Name, LOBELIA. *English Name*, INDIAN TOBACCO.

Definition.—The *leaves* and *tops* of Lobelia inflata.

Natural Order.—Lobeliaceæ. *Habitat.*—United States.

Contains the *liquid* alkaloid **lobeline** (not officinal) and lobelic acid. .

ESSENTIALS OF MATERIA MEDICA. 161

Officinal Preparations.
Extractum Lobeliæ Fluidum, ... ♏v-xv.
Tinctura Lobeliæ, ... ♏v-xxx.

Officinal Name, GELSEMIUM. *English Name*, YELLOW, OR CAROLINA, JASMINE.

Definition.—*Rhizome* and *rootlets* of Gelsemium sempervirens.

Natural Order.—Loganiaceæ. *Habitat.*—Southern United States.

Contains the alkaloid **gelsemine**, and gelseminic acid (non-officinal).

Officinal Preparations.
Extractum Gelsemii Fluidum, ... ♏iij-v.
Tinctura Gelsemii, ... ♏x-xxx.

Officinal Name, ACIDUM HYDROBROMICUM DILUTUM. *English Name*, DILUTED HYDROBROMIC ACID.

Definition.—Ten per cent. aqueous solution of absolute hydrobromic acid.

Dose, f℥ss-ij.

Officinal Name, TABACUM. *English Name*, TOBACCO.

Definition.—The commercial, dried *leaves* of Nicotiana tabacum.

Natural Order.—Solanaceæ. *Habitat.*—United States.

Contains the *liquid volatile* alkaloid **nicotine** and a volatile oil. Rarely used. There are no officinal preparations.

162 FOR THERAPEUTIC NOTES.

ESSENTIALS OF MATERIA MEDICA. 163

Officinal Name, CONIUM. *English Name*, HEMLOCK.

Definition.—The full-grown *fruit* of Conium maculatum or hemlock gathered while yet green. Its activity depends on **conine**, a *volatile liquid* alkaloid (not officinal).

Natural Order.—Umbelliferæ. *Habitat.*—United States and Europe.

Officinal Preparations.

Extractum Conii, ... gr. j-ij.
Extractum Conii Fluidum, ... ♏j-v.

As the preparations vary greatly in strength, begin always with the minimum dose and increase to desired effect.

FAMILY VII.—RESPIRATORY STIMULANTS.

Although classified under different headings, ammonia, caffeine, atropine, cocaine, and strychnine are valuable also as respiratory stimulants.

Officinal Name, ASPIDOSPERMA. *English Name*, QUEBRACHO.

Definition.—The *bark* of Aspidosperma quebracho-blancho (Schlechtendal).

Natural Order.—Apocynaceæ. *Habitat.*—South America.

Officinal Preparation.

Extractum Aspidospermatis Fluidum, . f℥ ¼-½.

Contains the alkaloid **aspidospermine**.
Dose, gr. ¼-ss.

..degree Ammonium 10% .. gas
9 ..

" 28% .. ga

..coholic Sol. of Amm-
..um Carb... Oil of La-
..vender 10 to 30 drops

ORDER II.—CARDIANTS.

FAMILY I.—CARDIAC STIMULANTS.

These remedies increase the force of the circulation, either by stimulating the heart muscle directly or by lessening resistance to the flow of blood; *i. e.*, dilating the arteries through which it runs.

AMMONIA.*

Definition.—A colorless irritant gas, alkaline in reaction, characteristic in odor, extremely soluble in water. Naturally found as a result of decaying animal and vegetable matter. Commercially obtained as a by-product in the manufacture of coal-gas.

Officinal Preparations.

Spiritus Ammoniæ, ... ♏v–f♍ ss.
Spiritus Ammoniæ Aromaticus, . . f♍ ss–ij.
Aqua Ammoniæ, ten per cent. gas, . . ♏x–xx.
Aqua Ammoniæ Fortior, 28 per cent. gas, ... ♏ij–x.
Liquor Ammonii Acetatis (dilute acetic acid neutralized by carbonate of ammonium—*Spirits of Mindererus*), . f♍ ss–ij.
Linimentum Ammoniæ.
Ammonii Benzoas, ... ⎫
Ammonii Bromidum, ... ⎬ gr. x–xxx.
Ammonii Carbonas, ... gr. ij–x.
Ammonii Chloridum, ... gr. v–♍ ss.

---

* Not officinal as gas.

*Dies on Colcothar(?)
the U.S.P. 1890.*

ESSENTIALS OF MATERIA MEDICA. 167

Trochisci Ammonii Chloridi, . . aa gr. $\frac{1}{10}$.
Ammonii Iodidum, ... gr. ij–x.
Ammonii Phosphas, ... gr. x–xv.
Ammonii Valerianas, ... gr. ij–x.

Officinal Name, ALCOHOL.

Definition.—A transparent volatile liquid; characteristic odor and taste; 91 per cent., by weight, of ethyl alcohol, and nine per cent., by weight, of water; specific gravity, 0.820.

Officinal Preparations.
Alcohol Dilutum.
Alcohol Absolutum, specific gravity, 0.797 ; contains only one per cent. water.
Alcohol Deodoratum, specific gravity, 0.816 ; 7½ per cent. water.

Officinal Name, ALCOHOL DILUTUM. *English Name*, DILUTED ALCOHOL.

Definition.—A liquid composed of 41 per cent., by weight, or about 48.6 per cent., by volume, of absolute ethyl alcohol, or about 59 per cent., by weight, of water. Alcohol exists in the following *officinal preparations*, viz. :

Spiritus Frumenti.—*Whisky*, 44 to 50 per cent. alcohol, by weight. Obtained by the distillation of fermented grains (practically distilled beer). Must be *at least two* years old.
Spiritus Vini Gallici.—*Brandy*, 39 to 47 per cent. alcohol, by weight. Obtained by distillation of fermented grapes (practically distilled wine). Must be *at least four* years old.

FOR THERAPEUTIC NOTES.

ESSENTIALS OF MATERIA MEDICA. 169

Vinum Album.—*White wine*, 10 to 12 per cent. alcohol.
Vinum Rubrum.—*Red wine*, 10 to 12 per cent. alcohol.
Vinum Album Fortius.—Twenty to 25 per cent. alcohol, used as menstruum for officinal wines.
Spiritus Odoratus.—*Cologne water.*
Vinum Aromaticum.—*Aromatic wine.*
Beer and the various malts (extracts, etc.) vary from three to eight per cent. of alcohol.

Officinal Name, DIGITALIS. *English Name*, FOXGLOVE (PURPLE).

Definition.—*Leaves* of Digitalis purpurea from plants of *second year's* growth. Contains four glucosides, viz.: *digitalin, digitoxin, digitonin,* and *digitalein.*
Natural Order.—Scrophularineæ. *Habitat.*—Europe.
Powdered leaves may be used in doses of gr. ss–iij.

Officinal Preparations.
Extractum Digitalis, ... gr. ¼–j.
Extractum Digitalis Fluidum, ... ♏ss–ij.
Tinctura Digitalis, ... ♏v–xv.
Infusum Digitalis, ... f℥j–v.

Officinal Name, CAFFEINA. *English Name*, CAFFEINE.
Definition.—Feeble alkaloid obtained from *dried leaves* of Thea sinensis (*natural order*, Ternstrœmiaceæ), ordinary tea, from Coffea arabica (*natural order*, Rubiaceæ), and from guarana* (a paste from

---
* Officinal.

crushed seeds of Paullinia sorbilis, *natural order*, Sapindaceæ).

Officinal Preparations.

Caffeina, ... gr. ij–x.
Caffeina Citrata, ... gr. x–xv.
Caffeina Citrata Effervescens, ... gr. x–xv.

Officinal Name, CONVALLARIA. *English Name*, LILY OF THE VALLEY.

Definition.—The *rhizome* and *rootlets* of Convallaria majalis, or lily of the valley.

Natural Order. — Liliaceæ. *Habitat.* — United States, Holland.

Contains the glucosides *convallarin* and *convallamarin*.

Officinal Preparation.

Extractum Convallariæ Fluidum, ... ♏v–xv.

Officinal Name, STROPHANTHUS. *English Name*, STROPHANTHUS.

Definition.—*Seeds* of Strophanthus hispidus deprived of its long awn. Used as arrow-poison by the African natives.

Natural Order.—Apocynaceæ. *Habitat.*—Africa.

Contains the active principle *strophanthin*.

Officinal Preparation.

Tinctura Strophanthi, ... ♏j–x.

.

.

.

~~Nitric acid~~
is given with Ethereal
or water injections of L

Officinal Name, SPARTEINÆ SULPHAS. *English Name,* SPARTEINE SULPHATE.

Definition.—The neutral sulphate of an alkaloid obtained from Scoparius or broom plant.

Natural Order.—Leguminosæ. *Habitat.*—Indigenous.

Dose, gr. $\frac{1}{20}-\frac{1}{4}$ hypodermically; gr. $\frac{1}{4}$–ij in pill.

ADONIDINE.*

Definition.—A *glucoside* from *root* of Adonis vernalis.

Natural Order. — Ranunculaceæ. *Habitat.* — Europe.

Dose, gr. ⅛–ss.

FAMILY II.—CARDIAC DEPRESSANTS.

These remedies depress the heart's action, and so reduce the force and frequency of the pulse.

ANTIMONIUM.†

Antimony (metallic element) is found as black antimonious sulphide.

Officinal Preparations.

Antimonii et Potassii Tartras (*Tartar Emetic*). Dose, as diaphoretic and expectorant, gr. $\frac{1}{12}-\frac{1}{8}$; as emetic, gr. ss–j, repeat if necessary.

---

* Not officinal.

† The metal itself is not officinal. Antidote, tannic acid.

(Antimony Sulphuratum)
mixture Pure Sulphide of An-
tide of antimony
diaphoretic Antimony not
officinal

———————

—ery active + overdose is likel-
ey to prove fatal —
May be confused with white
hellebore —

Tinct Veratrum Viridis 48% —
Dose small, 1 to 5 or 6 drop.
Antidote
?   ...   vomiting, keep

ESSENTIALS OF MATERIA MEDICA. 175

Pulvis Antimonialis (antimony oxide, 33
per cent. ; precipitated calcium phos-
phate, 67 per cent.), *James' powder*, . gr. iij-x.
Antimonii Sulphidum Purificatum.
Antimonum Sulphuratum, ... gr. j-v.
Pilulæ Antimonii Compositæ (*Plum-*   *1 gr Calom*
mer's pills), aa gr. ⅔ of antimony
and calomel.
Antimonium Oxidum, ... gr. j-ij.
Vinum Antimonii, four grams of tartar
emetic to 1000 c. c. of solution.
Diaphoretic and expectorant dose, . ♏v-xx.
Emetic dose, ... f ℥j-iv.
Syrupus Scillæ Compositus contains of
tartar emetic 2 gm. to 1000 c. c., . . ♏v-f℥j.

Officinal Name, VERATRUM VIRIDE. *English Name*,
. GREEN OR AMERICAN HELLEBORE.
Definition.—*Rhizome* and *rootlets* of Veratrum
viride.†
Natural Order.—Liliaceæ. *Habitat.*—Swamps of
Southern United States.
Alkaloids.—Jervine* and veratroidine.* *veratr*

Officinal Preparations.

{ Extractum Veratri Viridis Fluidum, . . ♏j-iv. } *1 tr 2 m.*
{ Tinctura Veratri Viridis, ... ♏ij-vj. }

---

* Not officinal.

† *Norwood's tincture* is made from the green root; and while
not officinal is more powerful than the officinal Tincture of
Veratrum Viride.

+ plenty of hot water – Oli
en rectum + genl. stimulant

Teratrine slightly sol in H2O
sol. in alcohol. An acute
ritant –
Dilate a 2% prep. Oleine &
Unguent. is 4% – Benzoated
Poison

Arnica Flowers & Root.
Til. Alcohol. to make Tinc. An.
Root.

x. Arnicine

Officinal Name, VERATRINA. *English Name,*
VERATRINE. white

Definition.—A mixture of alkaloids obtained from the *seed* of Asagræa officinalis.
Natural Order.—Liliaceæ. *Habitat.*—Mexico. S.A
Used externally for rheumatic pains.

Officinal Preparations.

Oleatum Veratrinæ, two per cent.
Unguentum Veratrinæ, four per cent.

Officinal Name,	*English Name,*
ARNICÆ FLORES.	ARNICA FLOWERS.
ARNICÆ RADIX.	ARNICA ROOT.

Definition.—The *flower-heads, rhizome* and *rootlets* of Arnica montana (leopard's bane).
Natural Order.—Compositæ. *Habitat.*—United States.

Officinal Preparations.

Flowers:
   Tinctura Arnicæ Florum (use externally). 20 % weak

Root:
   Extractum Arnicæ Radicis, ... gr. v–x.
   Extractum Arnicæ Radicis Fluidum, . ♏x–xx.
   Tinctura Arnicæ Radicis, ... f℥ss–ij. 10 % strong
   Emplastrum Arnicæ, 33 per cent. extract.

Officinal Name, ACONITUM. *English Name,* ACONITE.

Definition.—The *tuber* of Aconitum napellus, (monkshood or wolf's bane).
Natural Order. — Ranunculaceæ. *Habitat.*—Europe and Asia.

Antidote — Same as very me
to others.
Tinc. Aconite Leaves 10%
"   "   Root very strong.
Almost as strong as Veratrum.
Nearly all dil. acids are 10
we. HCN @ 7%.

Ammonia + stimulation of
all kinds. H cell made &
aleinizing the HCell.
But not a very stable cell
can be made from AgCN.
———————
V. eg. acids.
Tartaric acid

ESSENTIALS OF MATERIA MEDICA. 179

The alkaloid, **Aconitine,*** is almost too strong for internal use.

Dose, gr. $\tfrac{1}{200}$–$\tfrac{1}{500}$. Used mostly in liniments or ointments.

Officinal Preparations.

Extractum Aconiti, ... gr. ⅙–ss.
Extractum Aconiti Fluidum, ... ♏ss–ij.
Tinctura Aconiti, ... ♏j–v.

Officinal Name, ACIDUM HYDROCYANICUM DILUTUM.
English Name, DILUTED HYDROCYANIC ACID.

Definition.—Two per cent. solution in water of absolute hydrocyanic acid. Odor and taste those of peach kernels or bitter almonds.

Dose, ♏j–iij.

VEGETABLE ACIDS.

Officinal Name, ACIDUM TARTARICUM. *English Name*,
TARTARIC ACID.

Definition.—An organic acid usually prepared from argols (sediment of wine); colorless, translucent crystals, sour in taste.

Used only as an ingredient in Seidlitz powder, pulvis effervescens compositus, United States Pharmacopeia.

Officinal Name, ACIDUM CITRICUM. *English Name*,
CITRIC ACID.

Definition.—An organic acid prepared from

* Not officinal.

$H_2C_4H_4O_6$ Tartaric Acid
May distinguish Tartaric & Citric acids by behavior to R.
(Express juice & concentrate by evap.

Acetic acid — w him alcohol is ol. to ferment —

Pyroligneous acid — when wood is distilled — treat with Ca & pyrolignate of lime & treat ? & decompose & then get pure acetic acid.

Tests of Acetic acid.
$FeCl_2$ — Blood red color.

lemon- or lime-juice. Colorless, translucent, right-rhombic prisms; odorless; agreeable, acid taste. Soluble in $\frac{6}{10}$ of its weight of cold water, and in $\frac{6}{20}$ of its weight of boiling water. *Orange + orange*

Officinal Preparation.
Syrupus Acidi Citrici (8 : 1000), . . . f℥j–iv.

Officinal Name, ACIDUM ACETICUM. *English Name*,
ACETIC ACID. $HC_2H_3O_2$
Definition.—Colorless liquid, composed of 36 per cent., by weight, of absolute acetic acid, and 64 per cent. of water.
Used externally as a mild caustic.

Officinal Name, ACIDUM ACETICUM GLACIALE.
English Name, GLACIAL ACETIC ACID.
Definition.—Nearly or quite absolute acetic acid. Never used internally. *99% pure only, used h*

Officinal Name, ACIDUM ACETICUM DILUTUM. *English Name*, DILUTED ACETIC ACID.
Definition.—Six per cent., by weight, of absolute acetic acid ; water to make 100.
Dose, f℥j–ij, well diluted.
Used as menstruum for the officinal aceti.

Officinal Name, ACIDUM OXALICUM. *English Name*,
OXALIC ACID. $H_2C_2O_4 + 2H_2O$
Found naturally in sorrels and in other vegetable life.
Of interest only as a poison.

antidote — alkali — or chalk

Antidote.—Lime, calcium carbonate.
Antidote for all the above acids, except oxalic acid, is a mild alkali, followed by oil to prevent, if possible, further corrosion of the intestinal tract. Care must be taken in all cases not to give so much of the antidote that it may itself become poisonous in turn,—hence ammonia, unless very dilute, is contraindicated.

ORDER III.—NUTRIENTS.

FAMILY I.—ASTRINGENTS.

Definition.—Astringents are drugs which cause contractions of those tissues with which they are brought in contact, either directly or through the circulation.

Astringents are (1) vegetable and (2) mineral.

VEGETABLE ASTRINGENTS.

The vegetable astringents all depend for their action on tannic acid.

Officinal Name, ACIDUM TANNICUM. *English Name*, TANNIC ACID.

Definition.—An organic acid obtained from nut-galls. It is found widely distributed throughout the vegetable kingdom under two forms : (1) *gallo-tannic* (the officinal form); (2) *kino-tannic*. The gallo-tannic strikes a *blue-black* color with the salts of

$H C_{14} H_9 O_9$

$H C_7 H_5 O_5 + H_2 O = $ Gallic

$C_6 H_3 (OH)_3 = $ ~ ~~ash~~

Tannic acid coag. album.
blood + Gelatinized Starch
differs from Gallic acid.
Collodium = sol. gun-
cotton + Ether

iron, while kino-tannic gives a *greenish-black* precipitate.

It is a light, yellowish, amorphous powder, occurring in scales or masses; possessing a characteristic odor, and strong astringent taste. It coagulates albumen and is the chemic antidote for all the *alkaloids;* forming tannates which are very slow of solution by the intestinal tract. It forms an insoluble tannate of antimony, hence it is also the antidote for antimony in overdose.

Dose, as astringent, gr. iij–v ; as hemostatic, gr. x–xx.

Officinal Preparations.

Collodium Stypticum,	20 per cent.
Glyceritum Acidi Tannici,	20 per cent.
Trochisci Acidi Tannici,	1 = gr. j.
Unguentum Acidi Tannici,	20 per cent.

Officinal Name, ACIDUM GALLICUM. *English Name,* GALLIC ACID.

Definition.— An organic acid prepared from tannic acid by adding one molecule of water of crystallization. Does not coagulate albumen ; occurs in whitish, silky needles or triclinic prisms ; odorless, astringent taste. Does not precipitate alkaloids as does tannic acid.

Officinal Name, GALLA. *English Name,* NUTGALL.

Definition.—An excrescence on Quercus lusitanica (*natural order,* Cupuliferæ), caused by punctures and deposition of ova of Cynips gallæ tinctoriæ (*class,*

13

"Mercurius"

Itch in.
on the powdered

Tinctura Cd Catechu -
cinnamon.

Insecta; *order*, Hymenoptera). The best galls come from the Levant, and are usually the size of hickory-nuts.
Rarely used internally.

Officinal Preparations.
Tinctura Gallæ, ... f ℥ ss–ij.
Unguentum Gallæ, ... ten per cent.

Officinal Name, CATECHU. *English Name*, CATECHU.
Definition.—An *extract* from the wood of a tree —the Acacia catechu.
Natural Order.—Leguminosæ. *Habitat.*— East India.
Contains tannic acid.

Officinal Preparations.
Tinctura Catechu Composita, ... f ℥ ss–ij.
Trochisci Catechu, ... āā gr. j.

Officinal Name, KINO. *English Name*, KINO.
Definition.—The *inspissated juice* of Pterocarpus marsupium.
Natural Order.—Leguminosæ. *Habitat.*—East Indies.
Occurs in small, dark, brownish-red, shiny pieces; colors the saliva deep red. Contains kino-tannic acid.
Dose, gr. x–xx.

Officinal Preparation.
Tinctura Kino, ... f ℥ ss–ij.

Haematoxylon – pale yell-
ish needles. + Nitr make
d' tactile acid makes

Cram.
Peats.

Oaks 75 or so variety

ESSENTIALS OF MATERIA MEDICA. 189

Officinal Name, HÆMATOXYLON. *English Name*, LOG-WOOD.

Definition.—The *heart-wood* of Hæmatoxylon campechianum.

Natural Order.—Leguminosæ. *Habitat.*—Central America.

Contains hematin and tannic acid.

Gives a blue color in alkaline solution—red in acid.

Officinal Preparation.

Extractum Hæmatoxyli, ... gr. x–xxx.

Officinal Name, KRAMERIA. *English Name*, RHATANY.

Definition.—The *root* of Krameria triandra and Krameria ixina.

Natural Order.—Polygaleæ. *Habitat.*—South America.

Officinal Preparations.

Extractum Krameriæ, ... gr. v–x.
Extractum Krameriæ Fluidum, . . . ♏v–xxx.
Tinctura Krameriæ, ... f℥ ss–ij.

Officinal Name, QUERCUS ALBA. *English Name*, WHITE OAK.

Definition.—The *inner bark* of Quercus alba.

Natural Order.—Cupuliferæ. *Habitat.*—United States.

Used only for astringent washes and lotions.

Officinal Name, ROSA CENTIFOLIA. *English Name*, PALE ROSE.

Definition.—The *petals* of Rosa centifolia.

Raspberry — Syrupus Rubus occidentalis. Allow to ferment in order [to de]stroy the ?

ESSENTIALS OF MATERIA MEDICA. 191

Natural Order.—Rosaceæ.
Used only for its odor.

Officinal Name, ROSA GALLICA. *English Name*, RED ROSE.

Definition.—The *petals* of Rosa gallica collected before expanding.
Natural Order.—Rosaceæ.

Officinal Preparations.

Confectio Rosæ, ...
Extractum Rosæ Fluidum, ...
Confectio Rosæ, ... } vehicles.
Mel Rosæ, ...
Syrupus Rosæ, ...
Pilulæ Aloes et Mastiches.

Oil of rose is obtained from the *fresh flowers* of Rosa damascena (*natural order*, Rosaceæ). This is used in the preparation of **Aqua Rosæ** and **Unguentum Aquæ Rosæ** (cold cream).

Officinal Name, GERANIUM. *English Name*, CRANES-BILL, SPOTTED GERANIUM, ETC.

Definition.—The *rhizome* of Geranium maculatum.
Natural Order.—Geraniaceæ. *Habitat.*—United States. Contains large amount of tannic acid.

Officinal Preparation.

Extractum Geranii Fluidum, ... f ℥ ss–j.

does not depend so much
?amic acid as on Mattic a

UP [illegible sketches]

Lycopodium
    Devils snuff boxes.
_____
Alumens always have 2 mol. q
+ a regular octahedron
obtained: shake wet can
tain Al + from there we
it alum___.
Sol. in 9 parts q Cold
  "   " 1/3 q 1 " " Hot.

ESSENTIALS OF MATERIA MEDICA.

Officinal Name, RHUS GLABRA. *English Name*, RHUS GLABRA, SUMAC.

Definition.—The *fruit* of Rhus glabra.
Natural Order.—Anacardieæ. *Habitat.*—United States.

Contains tannic and malic acid.

Officinal Preparation.

Extractum Rhois Glabræ Fluidum (used as a gargle diluted with water).

MINERAL ASTRINGENTS.

Officinal Name, ALUMEN. *English Name*, ALUM.

Definition.—The double sulphate of aluminium and potassium. Soluble in nine parts of water at 59°, and in $\frac{3}{10}$ of a part of boiling water. Gives an acid reaction with litmus.

Dose, astringent, gr. x–xx; emetic, ʒj–iv; in lead poisoning, gr. xx–xl. Also used locally in solution as astringent and styptic.

Officinal Preparation.

Alumen Exsiccatum, sometimes called burnt or dried alum (used externally exclusively).

♄ PLUMBUM. LEAD.*

Found naturally as galena—lead ore, lead sulphide.

---

* Not official.

Alum incompat. with
Lead cels. alkalies & [alkali]
kaline carbonates,
tht is insol in carbonate alkali
KHO. would do same.
Bromo Chloralum.
    Lead —
Pb. not officinal. Water +
water plus NaCl + KNO₃ have
olvent effect. When SO₄ o or
eso though lead — form layers
upp on surface + are not so
 . protect them.
Saints have used on basis — Lead h
Due actite *, Poisoning — Anti
Chronic
H₂SO₄ — PbSo₄ utterly ins

Officinal Preparations.

For internal use :
  Plumbi Acetas (sugar of lead), ... gr. ss–v.

For external use :
  Liquor Plumbi Subacetatis (*Goulard's* extract).
  Liquor Plumbi Subacetatis Dilutus (lead water), three per cent. of Goulard's extract.
  Ceratum Plumbi Subacetatis, 20 per cent. (Goulard's cerate).
  Plumbi Carbonas (white lead).
  Plumbi Iodidum.
  Plumbi Nitras.
  Plumbi Oxidum (red lead, litharge). P&O.
  Unguentum Plumbi Carbonatis, . . . ten per cent.
  Unguentum Plumbi Iodidi, ... ten per cent.
  Emplastrum Plumbi.
  Unguentum Diachylon (lead plaster, 500; olive oil, 490; oil of lavender flowers, 10).

Poisoning may occur from any of the lead preparations.

Antidote.—Soluble sulphate or dilute sulphuric acid; alum in chronic poisoning; dilute sulphuric acid as drink, and iodide of potassium.

BISMUTHUM. BISMUTH.*
Metallic element.

---

* Not officinal.

ESSENTIALS OF MATERIA MEDICA. 197

Officinal Preparations.
Bismuthi Citras, ... gr. j–iij.
Bismuthi et Ammonii Citras, ... gr. j–v.
Bismuthi Subnitras and Bismuthi Subcarbonas, gr. v–xv in affections of stomach, and gr. xv–ʒj in affections of intestines.

Officinal Name, CERII OXALAS. *English Name*,
CERIUM OXALATE.

Definition.—A white, granular powder; odorless; tasteless. Insoluble in water, alcohol, or ether, but soluble in sulphuric acid. Used principally to relieve the vomiting of pregnancy.

Dose, gr. j–iij, in pill t. i. d.

ZINCUM. ZINC.

A metal obtained in the form of carbonates and sulphides. Officinal in the form of thin sheets, pencils or fine powder.

Officinal Preparations.
Zinci Oxidum, ... gr. j–v.
Unguentum Zinci Oxidi, ... 20 per cent.
Zinci Acetas (used in solution—gr. j–v–xx to f ʒ —as eye-wash or injection in gonorrhea).
Zinci Acetas, ... gr. j–v.
Zinci Bromidum, ... gr. j–v.
Zinci Carbonas Præcipitatus, ... external use.
Zinci Chloridum, ... external use.
Liquor Zinci Chloridi, ... external use.
Zinci Iodidum, ... gr. j–v.

☉ = Sol = Aurum
☾ = Luna = Argentum = Ag
♄ = Saturnus = Plumb
♃ = Jupiter = Stannum
♀ = Venus = Cuprum
♂ = Mars = Ferrum

Zinci Phosphidum, ... gr. j–v.
Zinci Sulphas (*White Vitriol*). **Dose,**
astringent, gr. j–ij ; emetic, gr. xxx,
repeated if required.

Antidote.—Alkalies and alkaline carbonates.
Albumen to be used as demulcent.

, CUPRUM. COPPER.* = ♀ *Venus*

Officinal Preparations.

Cupri Sulphas (*Blue Vitriol*). Dose,
astringent, gr. ¼–j ; emetic, ... gr. iij–v.
Cupri Acetas,* . . . . gr. ⅛–¼.

Copper sulphate is the antidote for poison by
phosphorus.

Antidote.—Potassium ferrocyanide, albumens as
demulcents.

/ ARGENTUM. SILVER.* ☾ = *Luna*

Officinal Preparations.

Argenti Cyanidum (pharmacy).
Argenti Iodidum, ... gr. ss–j in pill.
Argenti Nitras, ... gr. ⅛–ss in pill.
Solutions of different strengths are used
as eye-washes, injections, etc.
Argenti Nitras Dilutus (one-half each,
silver and potassium nitrate).
Argenti Nitras Fusus (*Lunar Caustic*,
external use).
Argenti Oxidum, ... gr. ss–j in pill.

* Not officinal.

igO, loosely held & easily
se the O & cause an expo

Family II Tonics.

ore important.

Owing to the readiness with which silver decomposes it should invariably be given in pill, freshly prepared, when intended for internal administration.

Antidote.—Sodium chloride (common salt) forms an insoluble chloride.

FAMILY II.—TONICS.

Tonics are remedies employed to increase the strength and vigor of the body or its organs when depressed from disease or other causes.

MINERAL TONICS.

Officinal Name, FERRUM.   *English Name*, IRON.*

Officinal Preparations.

Ferri Carbonas Saccharatus, ... gr. v–ʒ ss.
Ferri Chloridum.
Ferri Citras.
Ferri et Ammonii Citras, ... gr. ij–v.
Ferri et Ammonii Sulphas, ... gr. ij–v.
Ferri et Ammonii Tartras, ... gr. j–x.
Ferri et Potassii Tartras, ... gr. v–x.
Ferri et Quininæ Citras, ... gr. v–x.
Ferri et Quininæ Citras Solubilis.
Ferri et Strychninæ Citras, one per cent., ... gr. j–v.
Ferri Hypophosphis, ... gr. v–x.
Ferri Iodidum Saccharatum, ... gr. v–xv.
Ferri Lactas, ... gr. v–x.
Ferri Oxidum Hydratum.

---

* Officinal in the form of fine, bright, and non-elastic wire.

Ferri Oxidum Hydratum cum Magnesia (used in *arsenical poisoning*).
Ferri Phosphas Solubilis, ... gr. ij–v.
Ferri Pyrophosphas Solubilis, ... gr. ij–v.
Ferri Sulphas (*green vitriol*), . . .
Ferri Sulphas Exsiccatus, ... } gr. ss–ij.
Ferri Sulphas Granulatus, ...
Ferri Valerianas, ... gr. j–iij.

Besides the above, the most commonly used preparations of iron are:

Ferrum Reductum (reduced iron, *Quevenne's iron*), ... gr. ij–v in pill.
Massa Ferri Carbonatis (*Vallet's mass*), gr. iij–x in pill.
Liquor Ferri Tersulphatis (used with Ferri Oxidum Hydratum and Ferri Oxidum Hydratum cum Magnesia, as antidotes in arsenical poisoning).
Liquor Ferri Subsulphatis (*Monsel's solution*) used locally as styptic.
Tinctura Ferri Chloridi, ... ♏v–xxx.
Syrupus Ferri Iodidi, ... ♏v–xxx.
Liquor* Ferri et Ammonii Acetatis (*Basham's mixture*, tonic and diuretic), ... f ʒ j–iv.
Ferrum Dialysatum,† . . . . f ʒ ss–f ʒ j.
As antidote for arsenic, ... f ℥ ss.
Liquor Ferri Acetatis, ... external use.
Liquor Ferri Chloridi, ... external use.
Liquor Ferri Citratis, ... ♏v–xv.
Liquor Ferri Nitratis, ... ♏ij–x.
Syrupus Ferri Iodidi, ... f ʒ j–ij.

* Mistura (Pharm., 1880). † Not officinal.

Syrupus Ferri, Quininæ, et Strychninæ
  Phosphatum, ... f ℥ j–ij.
Vinum Ferri Amarum, ... f ℥ j–ij.
Vinum Ferri Citratis, ... f ℥ j–ij.
Pilulæ Aloes et Ferri, ... j pill.
Pilulæ Ferri Carbonatis, ... j–iij pills.
Pilulæ Ferri Iodidi, ... j–iij pills.
Trochisci Ferri (1 = gr. v of ferric hy-
  drate), ... j–v troches.
Emplastrum Ferri.

MANGANUM. MANGANESE.*

Officinal Preparations.
Mangani Oxidum Nigrum (1880), . . gr. j–x in pill.
Mangani Sulphas, ... gr. j–v.
Mangani Dioxidum (same as Mangani
  Oxidum Nigrum, of Pharm., 1880).

Officinal Name, ACIDUM SULPHURICUM (H_2SO_4).
English Name, SULPHURIC ACID (called sometimes OIL OF VITRIOL).

Definition.—A liquid composed of not less than 92.5 per cent., by weight, of absolute sulphuric acid, and not more than 7.5 per cent. of water. Used internally only if diluted. Poisoning causes charring and blackening of the tissues.

Officinal Preparations.
Acidum Sulphuricum Dilutum (ten per
  cent. of officinal acid), ... ♏v–xx further di-
Acid Sulphuricum Aromaticum (*elixir*   [luted.
  of vitriol), ten per cent., ... ♏v–xx.

---

* Not officinal.

Officinal Name, ACIDUM HYDROCHLORICUM (HCl).
English Name, HYDROCHLORIC OR MURIATIC
ACID.

Definition.—A colorless liquid, composed of 32 per cent., by weight, of absolute hydrochloric acid.
Not used internally. Poisoning leaves a yellow stain on the tissues.

Officinal Preparation.
Acidum Hydrochloricum Dilutum (ten
per cent. of absolute acid), . . ℳx–xx.

Officinal Name, ACIDUM NITRICUM (HNO_3). *English Name,* NITRIC ACID.

Definition.—A liquid composed of 68 per cent., by weight, of absolute nitric acid, and 32 per cent. of H_2O.
Not used internally. Poisoning causes deep orange-yellow staining of the tissues.

Officinal Preparation.
Acidum Nitricum Dilutum (ten per cent.
of absolute acid), ... ℳx–xx.

Officinal Name, ACIDUM NITROHYDROCHLORICUM.
English Name, NITROHYDROCHLORIC ACID.

Definition.—Consists of 18 parts nitric acid and 82 parts hydrochloric acid.

Officinal Preparation.
Acidum Nitrohydrochloricum Dilutum
(nitric acid, four parts; hydrochloric
acid, 18 parts; and water, 78 parts),'. ℳx–xx.

Officinal Name, ACIDUM LACTICUM. *English Name*, LACTIC ACID.

Definition.—The acid of milk-sugar, composed of 75 per cent. of absolute lactic acid.

Dose, ♏v–xxx.

Officinal Name, PHOSPHORUS. *English Name*, PHOSPHORUS.

Definition.—Translucent, almost colorless, solid; waxy in appearance and consistency. Has an odor resembling garlic, and characteristic taste (should never be tasted unless greatly diluted). Takes fire spontaneously on exposure to the air.

Dose, gr. $\frac{1}{100}$.

Officinal Preparations.
Oleum Phosphoratum (one per cent. of phosphorus in oil of almonds and ether), ... ♏j–iij.
Pilulæ Phosphori, ... aa gr. $\frac{1}{100}$.
Spiritus Phosphori (used to make elixir).
Zinci Phosphidum, ... gr. $\frac{1}{20}$–¼.
Elixir Phosphori (two per cent. of spirit), ♏x.

Antidotes.—Sulphate of copper, French oil of turpentine.

FAMILY III.—ALTERATIVES.

Medicines which in some way seem to alter the nutrition and increase the strength and health in various pathologic states.

ESSENTIALS OF MATERIA MEDICA. 211

ARSENUM. ARSENIC.*

Definition.—A metallic element found often with other metals as an arsenide. Black arsenic, sometimes called cobalt.

Officinal Preparations.

Acidum Arsenosum (white arsenic), . gr. $\frac{1}{20}$.
Liquor Acidi Arsenosi, ... ⎫
Liquor Potassii Arsenitis (*Fowler's solution*), ... ⎬ ♏ij–x.
Liquor Sodii Arsenatis—each of
the three above liquors contain
one per cent. of arsenous acid, . ⎭
Sodii Arsenas, ... gr. $\frac{1}{12}$–$\frac{1}{4}$.
Arseni Iodidum, ... gr. $\frac{1}{8}$.
Liquor Arseni et Hydrargyri Iodidi
(*Donovan's solution*, one per cent.
each of arsenic and mercuric iodide), ♏ij–x.

Antidote.—Freshly prepared hydrated oxide of iron with magnesia in large amount, emetics, etc.

Officinal Name, HYDRARGYRUM. *English Name*, MERCURY.

Definition.—A liquid element obtained from the sulphide cinnabar. Specific gravity, 13.5.

Officinal Preparations.
For internal use:
Hydrargyrum cum Creta (gray powder),
38 per cent. of mercury, small doses as
laxative, large doses as antisyphilitic, gr. j–v–xx.

---
* Not officinal.

For internal use :
Massa Hydrargyri (blue mass), 33 per
  cent. mercury, ... gr. j–x.
Hydrargyri Chloridum Corrosivum (bi-
  chloride, corrosive sublimate), . . . gr. $\frac{1}{30}-\frac{1}{10}$.
Hydrargyri Chloridum Mite (calomel), . gr. ss–xx.
Hydrargyri Cyanidum, ... gr. $\frac{1}{20}-\frac{1}{12}$.
Hydrargyri Iodidum Rubrum (red
  iodide), ... gr. $\frac{1}{30}-\frac{1}{10}$.
Hydrargyri Iodidum Viride (Pharm.,
  1880), same as Hydrargyri Iodidum
  Flavum, ... gr. $\frac{1}{4}$–j.
Hydrargyri Iodidum Flavum (prot-
  iodide, yellow or green iodide), . . gr. $\frac{1}{4}$–j.
Pilulæ Catharticæ Compositæ (see Colo-
  cynth), ... j–iij pills.
Hydrargyri Subsulphas Flavus (*turpeth
  mineral*), as emetic, ... gr. ij–v.
Pilulæ Antimonii Compositæ (*Plummer's
  pill*, calomel and sulphurated antimony).

Antidote.—Albumen (white of egg), milk, wheat flour, emetic.

For external use only :
Emplastrum Hydrargyri.
Emplastrum Ammoniaci cum Hydrargyri.
Unguentum Hydrargyri (blue ointment),
  45 per cent. mercury.
Unguentum Hydrargyri Ammoniati, ten per cent.
Unguentum Hydrargyri Oxidi Rubri, ten per cent.
Unguentum Hydrargyri Oxidi Flavi, ten per cent.
Unguentum Hydrargyri Nitratis (citrine ointment).
Hydrargyri Ammoniatum (white precipitate).

For external use only:
   Hydrargyri Oxidum Rubrum (red precipitate).
   Hydrargyri Oxidum Flavum.
   Oleatum Hydrargyri (yellow oxide), ten per cent.
   Liquor Hydrargyri Nitratis (caustic).
   Hydrargyri Subsulphidum Rubrum (cinnabar), used in fumigation.*

Officinal Name, AURI ET SODII CHLORIDUM. *English Name*, CHLORIDE OF GOLD AND SODIUM.

Definition.—A mixture of equal parts, by weight, of dry gold chloride and sodium chloride.

Dose, gr. $\frac{1}{12}$–$\frac{1}{4}$ in pill.

Officinal Name, IODUM. *English Name*, IODINE.

Definition.—Non-metallic element made from the ashes of seaweed; forms blue color with starch, which is the antidote in poisoning by it.

Officinal Preparations.

Iodi, ... gr. ¼–j.
Liquor Iodi Compositus (*Lugol's solution*), iodine, 5; potassium iodide, 10; water, 85, ... ♏ iij–x.
Ammonii Iodidum, ... gr. ij–x.
Potassii Iodidum, ... gr. v–ℨ ss.
Strontii Iodidum, ... gr. v–ℨ ss.
Syrupus Acidi Hydriodici, ... f ℨ ss–ij.

Externally only:
   Tincturæ Iodi, ... eight per cent.
   Unguentum Iodi, ... four per cent.
   Unguentum Potassii Iodidi, ... twelve per cent.

---

* Not officinal.

Officinal Name, IODOFORMUM. *English Name*,
IODOFORM.

Definition.—Small yellow crystals, characteristic odor, and iodine-like taste ; slightly soluble in water; soluble in alcohol, chloroform, and ether ; volatile. Used chiefly as an antiseptic.

Dose, gr. j–iv.

Officinal Preparation.
Unguentum Iodoformi, ... ten per cent.

Aristol and **iodol*** may be used in the same way as iodoform.

Officinal Name, OLEUM MORRHUÆ, OLEUM JECORIS ASELLI. *English Name*, COD-LIVER OIL.

Definition.—A fixed oil obtained from the fresh livers of Gadus morrhua and other species of Gadus. Contains gadium, iodine, chlorine, bromine, and fatty acids.

Dose, f₃j–f℥ss, t. i. d. ; mostly given in emulsion.

Officinal Name, ACIDUM PHOSPHORICUM. *English Name*, PHOSPHORIC ACID.

Definition.—A liquid, containing not less than 85 per cent., by weight, of absolute orthophosphoric acid. For external use and pharmacy only.

Officinal Preparation.
Acidum Phosphoricum Dilutum (ten per
cent. of absolute acid), ... ♏v–xxx.

---

* Not officinal.

ESSENTIALS OF MATERIA MEDICA. 219

Officinal Name,   English Name,
COLCHICI SEMEN.   COLCHICUM SEED.
COLCHICI RADIX.   COLCHICUM ROOT.

Definition.—The *seeds* and *corm* of Colchicum autumnale, or meadow saffron.

Natural Order.—Liliaceæ.   *Habitat.*—Europe.

Contains the alkaloid **colchicine**, which is its active principle.

Used in the form of the salicylate.

Dose, gr. $\frac{1}{50}$.

Seed: *Officinal Preparations.*
    Extractum Colchici Seminis Fluidum, . ♏ij–vi.
    Tinctura Colchici Seminis, ... f ℥ ss–j.
    Vinum Colchici Seminis, ... ♏v–xxx.

Root:
    Extractum Colch i Radicis, ... gr. ss–ij.
    Extractum Colch. Radicis Fluidum, . ♏ij–v.
    Vinum Colchici Radicis, ... ♏v–xv.

Officinal Name, SARSAPARILLA.   *English Name*, SARSA-
                                                                 PARILLA.

Definition.—The *root* of Smilax officinalis, Smilax medica, and other varieties of Smilax.

Natural Order.— Liliaceæ.   *Habitat.* — Mexico, South and Central America.

Contains the glucoside *smilacin*.

Officinal Preparations.
Decoctum Sarsaparillæ Compositum, . f ℥ ij–iv.
Extractum Sarsaparillæ Fluidum, . . . f ℥ j.
Extractum Sarsaparillæ Fluidum Com-
    positum, ... f ℥ j.
Syrupus Sarsaparillæ Compositus, . . f ℥ ss.

FOR THERAPEUTIC NOTES.

ESSENTIALS OF MATERIA MEDICA. 221

Officinal Name, *English Name*,
GUAIACI LIGNUM. GUAIACUM WOOD.
GUAIACI RESINA. GUAIAC.

Definition.—The *heart-wood* of Guaiacum officinale and Guaiacum sanctum, an evergreen tree of South America. The resin is obtained from the wood, and the preparations are made from it.

Natural Order.—Zygophylleæ.

Dose, gr. v–xx.

Officinal Preparations.
Tinctura Guaiaci, ... f ℥ ss–ij.
Tinctura Guaiaci Ammoniata, ... f ℥ ss–ij.

Officinal Name, MEZEREUM. *English Name*,
MEZEREUM.

Definition.—*Bark* of Daphne mezereum, and other species of Daphne.

Natural Order.—Thymelæaceæ. *Habitat.*—Europe.

Contains the neutral, bitter glucoside *daphnin*.

Officinal Preparation.
Extractum Mezerei Fluidum (used only for pharmaceutic purposes) is used in both the compound decoction and compound fluid extract of sarsaparilla.

Officinal Name, SASSAFRAS. *English Name*,
SASSAFRAS.

Definition.—The *bark* of the *root* of Sassafras variifolium.

Natural Order.—Laurineæ. *Habitat.*— Europe and United States.

Contains a volatile oil, used mostly for flavoring.

Officinal Name, TARAXACUM. *English Name*, DANDELION.

Definition.—The *root* of Taraxacum officinale.

Natural Order.—Compositæ. *Habitat.*—Indigenous.

Contains *taraxacin*, a bitter principle.

Officinal Preparations.
Extractum Taraxaci, ... gr. xx–ʒj.
Extractum Taraxaci Fluidum, ... f ʒj–ij.

ICHTHYOL.*

Definition.—A substance obtained by the distillation of a bituminous, sulphurous mineral deposit, found in North Germany, due to decomposition of fossil fish. Is now made synthetically. Contains about ten per cent. of sulphur.

Used externally as ointment; internally, gr. j–iij in pill or capsule.

FAMILY IV.—ANTIPERIODICS.

These remedies—which have for their type quinine—are used to overcome periodic fevers; *e. g.*, malaria.

Officinal Name, CINCHONA.

Definition.—The *bark* of any species of Cinchona which contains five per cent. of total alkaloids, and at least 2.5 per cent. of quinine ($C_{20}H_{24}N_2O_2 + H_2O$).

Natural Order.—Rubiaceæ. *Habitat.*—Peru and Bolivia.

---

* Not officinal.

The bark of **Cinchona Calisaya, Cinchona officinalis,**—*English name,* yellow cinchona,— is used in preparing the following:

Officinal Preparations.

Extractum Cinchonæ, ... gr. v–xx.
Extractum Cinchonæ Fluidum, . . . f℥j.
Tinctura Cinchonæ, ... f℥j–iv.
Infusum Cinchonæ, ... f℥j–ij.

Officinal Name, CINCHONA RUBRA. *English Name,* RED CINCHONA.

Definition.—The *bark* of Cinchona succirubra.
Natural Order.—Rubiaceæ.

Officinal Preparation.

Tinctura Cinchonæ Composita (Huxham's tincture), ... f℥j–iv.

The **alkaloids** of cinchona in the order of their potency are (1) quinine, (2) cinchonine, (3) quinidine, (4) cinchonidine.

Officinal Alkaloids and Salts.

Quinina, ...	As a tonic, gr. j–
Quininæ Sulphas, ...	iij ; as an anti-
Quininæ Bisulphas (more soluble than the sulphate), ...	pyretic, gr. v–xx ; as an anti-
Quininæ Hydrochloras, ...	periodic, gr. v–
Quininæ Hydrobromas, ...	xxx
Quininæ Valerianas, ...	gr. j–v.
Cinchonina, ...	Doses about one-
Cinchoninæ Sulphas, ...	third larger than
Cinchonidinæ Sulphas, ...	quinine.

Quinine and quinidine, or any salt of either, if treated with fresh chlorine or bromine water form an *emerald-green* precipitate if ammonia water is added to the solution. Cinchonine and cinchonidine, or their salts, form a *white* precipitate when thus treated.

Aqueous solutions of quinine, quinidine, and their salts, when acidulated with sulphuric acid, produce a pale *bluish* efflorescence. A weak solution of cinchonine or its salts should not exhibit more than a pale *yellow* color. Morphine imparts to sulphuric acid only a pale *yellow* tinge. Quinine and quinidine, and their salts, should not cause a *red* color with nitric acid as does morphine.

WARBURG'S TINCTURE.*

Definition.—A dark-brown liquid containing numerous ingredients much used in the pernicious malarial fevers of India. (See formula in United States Dispensatory.)

Dose, f℥ss.

Officinal Name, EUCALYPTUS. *English Name*, EUCALYPTUS.

Definition.—The *leaves* of Eucalyptus globulus collected from the older parts of the tree.

Natural Order.—Myrtaceæ. *Habitat.*—Australia.

Officinal Preparations.

Extractum Eucalypti Fluidum, ... f℥j–ij.
Oleum Eucalypti, ... ♏v–xx.

---
* Not officinal.

ACIDUM PICRICUM. PICRIC ACID.*

Trinitrophenol, $C_6H_2(NO_2)_3OH$.

Definition.—Made by dissolving crystals of carbolic acid in strong sulphuric acid, and adding nitric acid to the solution.

Used mostly in the arts.

FAMILY V.—ANTIPYRETICS.

Remedies which do not affect the normal temperature but cause a reduction of the temperature in febrile conditions.

Officinal Name, ACIDUM CARBOLICUM. *English Name*, CARBOLIC ACID.

Definition.—A constituent of coal-tar, obtained by distillation and then purified; also called *phenic* and *phenylic acid*.

The crude form is used as a disinfectant. Occurs in the form of needle-shaped crystals, white when pure but inclined to turn red.

Used internally in **doses** of gr. j.

Officinal Preparation.
Unguentum Acidi Carbolici, ten per cent.

Antidote.—Soluble sulphate (magnesium sulphate) rapidly administered in large amount.

---

* Not officinal.

Officinal Name, CREOSOTUM.   *English Name,*
CREOSOTE.

Definition.—A mixture of phenols, chiefly of guaiacol (the value of preparation depends on the amount of guaiacol it contains), and creosol, obtained by the distillation of wood-tar. That from beech-wood (Fagus sylvatica, *natural order,* Cupuliferæ) is preferred.

Dose, ♏j–iij.

Officinal Preparation.

Aqua Creosoti, one per cent., ... f℥j–iv.

GUAIACOL.*

Definition.—A syrupy liquid found in creosote; often used in the form of the carbonate.

Dose, gr. v.

Guaiacol itself is used in **doses** of five drops—♏xx–xl daily.

Officinal Name, MENTHOL.   *English Name,* MENTHOL.

Definition.—A stearopten obtained from the officinal oil of peppermint (Mentha piperita) or other menthæ.

Used as a local anesthetic by rubbing it on the part to be anesthetized.'

Officinal Name, THYMOL.

Definition.—A phenol obtained from the oil of thyme. It is a local anesthetic and antiseptic, and often used as a spray in throat and mouth affections.

---

* Not officinal.

FOR THERAPEUTIC NOTES.

ESSENTIALS OF MATERIA MEDICA. 233

Officinal Name, RESORCINUM. *English Name*,
RESORCIN.

Definition.—A diatomic phenol, antiseptic and antifermentative.

Dose, gr. ij–v.

Officinal Name, ACIDUM SALICYLICUM. *English Name*,
SALICYLIC ACID.

Definition.—An organic acid found in most plants, but generally prepared synthetically from carbolic acid. Soluble in 450 parts of water and in 2.4 of alcohol.

Dose, gr. x–ʒj.

Officinal Preparations.

Sodii Salicylas, ... gr. v–ʒ ss.
Lithii Salicylas, ... gr. j–x.
(ʒj–jss of the Sodii Salicylas in twenty-four hours, preferably administered in milk.)

Officinal Name, OLEUM GAULTHERIÆ. *English Name*,
OIL OF WINTERGREEN.

Definition.—A volatile oil distilled from the *leaves* of the Gaultheria procumbens.

Natural Order.—Ericaceæ. *Habitat.*—United States.

Consists almost entirely of *methyl salicylate*, to which its virtue is due.

Dose, ♏x–xv, in capsule or emulsion.

Officinal Preparation.

Spiritus Gaultheriæ, ... f ʒ ss–j.

16

ESSENTIALS OF MATERIA MEDICA.   235

Officinal Name, SALICINUM.   *English Name*, SALICIN.

Definition.—A neutral principle obtained from several species of Salix (willow). Consists of white, silky, crystalline needles; odorless, but of very bitter taste. Soluble in hot or cold water.

· **Dose,** gr. x–ʒj.

Officinal Name, BENZOINUM.   *English Name*, BENZOIN.

Definition.—A balsamic resin obtained from Styrax benzoin.

Natural Order.—Styraceæ.   *Habitat.*—Peru.

Contains *benzoic acid, a volatile oil and a resin.*

Officinal Preparations.

Adeps Benzoinatus, ... external use.
Tinctura Benzoini, ... fʒj–ij.
Tinctura Benzoini Composita, . . . fʒj–ij.

Officinal Name, SALOL.   *English Name*, SALOL.

Definition.—The salicylic ether of phenol. A white, crystalline powder resembling in odor oil of wintergreen; tasteless; almost insoluble in water.

Dose, gr. v–xv, t. i. d.

ANTIPYRIN.*

Definition.—A white, odorless powder, having a slightly bitter taste. Obtained synthetically and by the distillation of coal-tar.

Dose, gr. x–xx.

---

* Not officinal.

ANTIFEBRIN (ACETANILIDUM,* PHENACETIN). †
Definition.—Coal-tar products, all resembling antipyrin in appearance and effect.
Dose, gr. v–xv, repeated if necessary.

THALLIN. †
Definition.—A synthetically prepared alkaloid; antipyretic.
Dose, gr. v–x.

FAMILY I.—STOMACHICS.
Substances which, by increasing the activity of the glands of the gastro-intestinal tract, facilitate digestion.

The simple bitters are all of vegetable origin, of bitter taste, and while they stimulate markedly the mucous membrane of the gastro-intestinal tract, have practically no effect on the general system.

They include **quassia** and **calumba**, whose preparations contain no *tannic acid* and may therefore be used with iron preparations.

Almost if not all other vegetable preparations contain tannic acid, and are therefore incompatible with iron preparations.

SIMPLE BITTERS.

Officinal Name, QUASSIA.

Definition.—The *wood* of Picræna excelsa.

---

* Acetanilidum is officinal. † Not officinal.

ESSENTIALS OF MATERIA MEDICA. 239

Natural Order.—Simarubeæ. *Habitat.*—Jamaica.
Contains *quassin*, a neutral, bitter principle.

Officinal Preparations.

Extractum Quassiæ, ... gr. j–v.
Extractum Quassiæ Fluidum, ... ♏x–xxx.
Tinctura Quassiæ, ... f℥ ss–ij.

An infusion is often used as an enema in treatment for seat-worms.

Officinal Name, GENTIANA. *English Name*, GENTIAN.

Definition.—The *root* of Gentiana lutea (yellow alpine gentian).
Natural Order.—Gentianeæ. *Habitat.*—Europe.
Contains the active principle *gentiopikrin* and gentisic acid.

Officinal Preparations.

Extractum Gentianæ, ... gr. ij–x.
Extractum Gentianæ Fluidum, ... ♏x–xxx.
Tinctura Gentianæ Composita, . . . f℥j–iv.

Officinal Name, HYDRASTIS. *English Name*, GOLDEN SEAL.

Definition.—The *rhizome* and *rootlets* of Hydrastis canadensis, tumeric root, etc.
Natural Order.—Ranunculaceæ. *Habitat.*—Indigenous.
Contains the alkaloids **hydrastine** and **berberine**.

ESSENTIALS OF MATERIA MEDICA. 241

Officinal Preparations.
Extractum Hydrastis Fluidum, . . . ♏x–f℥j.
Glyceritum Hydrastis, ... f℥j.
Tinctura Hydrastis, ... f℥ss–ij.
Hydrastininæ Hydrochloras, ... gr. ¼.

Officinal Name, CALUMBA. *English Name*, COLUMBO.
Definition.—The *root* of Jateorhiza palmata.
Natural Order.—Menispermaceæ. *Habitat.*—Africa.
Contains the alkaloid **berberine**, which is found in many plants, and *columbin*, a bitter principle; but *no* tannic acid.

Officinal Preparations.
Extractum Calumbæ Fluidum, ... ♏xv–f℥ss.
Tinctura Calumbæ, ... f℥j–ij.

Officinal Name, EUPATORIUM. *English Name*, THOROUGHWORT, BONESET.
Definition.—The *leaves* and *flowering tops* of Eupatorium perfoliatum.
Natural Order.—Compositæ. *Habitat.*—Indigenous.

Officinal Preparation.
Extractum Eupatorii Fluidum, ... f℥j–iv.

Officinal Name, PRUNUS VIRGINIANA. *English Name*, WILD CHERRY BARK.
Definition.—The *inner bark* of Prunus serotina, collected in autumn.

Natural Order.—Rosaceæ. *Habitat.*—United States.

It contains the glucoside *amygdalin*, tannic acid, bitter extractives, and emulsin.

Officinal Preparations.
Extractum Pruni Virginianæ Fluidum, f ℨ ss.
Infusum Pruni Virginianæ, ... f ℥ ij.
Syrupus Pruni Virginianæ, ... vehicle.

Officinal Name, CHIRATA.

Definition.—The entire *plant,* Swertia chirata.
Natural Order.—Gentianeæ. *Habitat.*—India.

Officinal Preparations.
Extractum Chiratæ Fluidum, ... ♏ x–xxx.
Tinctura Chiratæ, ... f ℨ ss–j.

AROMATICS.

Used mostly as carminatives (to expel flatus), to disguise the taste of other medicines, and to prevent the griping of purgatives. They all contain a volatile oil, to which their activity is due.

Officinal Name, CINNAMOMUM ZEYLANICUM. *English Name,* CEYLON CINNAMON (CINNAMOMUM, Pharm., 1880).

Definition.—The *inner bark* of the shoots of Cinnamomum zeylanicum.
Natural Order.—Laurineæ. *Habitat.*—Ceylon.

ESSENTIALS OF MATERIA MEDICA. 245

Officinal Preparations.

Oleum Cinnamomi, ... ♏j–v.
Aqua Cinnamomi, ... vehicle.
Spiritus Cinnamomi, ... ♏x–f℥ss.
Tinctura Cinnamomi, ... f℥ ss–ij.
Pulvis Aromaticus, contains cinnamon,
 ginger, nutmeg, and cardamom, . . gr. x–xx.
Extractum Aromaticum Fluidum, . . ♏x–xx.

Officinal Name, MYRISTICA. *English Name,* NUTMEG.

Definition.—The *seed* of Myristica fragrans, deprived of its testa.

Natural Order.—Myristicaceæ. *Habitat.*—West Indies and various islands of Asia.

Used in pulvis aromaticus and tinctura lavandulæ composita.

Officinal Preparations.

Oleum Myristicæ, ... ♏ij–v.
Spiritus Myristicæ, ... f℥ ss–ij.

Officinal Name, CARYOPHYLLUS. *English Name,* CLOVES.

Definition.—The unexpanded *flowers* of Eugenia aromatica.

Natural Order.—Myrtaceæ. *Habitat.*—Molucca Islands, India.

Used in tinctura lavandulæ composita.

Officinal Preparation.

Oleum Caryophylli, ... ♏j–v.

ESSENTIALS OF MATERIA MEDICA. 247

Officinal Name, PIMENTA. *English Name*, ALLSPICE.
Definition.—The nearly *ripe fruit* of Pimenta officinalis.
Natural Order.—Myrtaceæ. *Habitat.*—West Indies.

Officinal Preparation.
Oleum Pimentæ, ... ♏j–v.

Officinal Name, CARDAMOMUM. *English Name*, CARDAMOM, OR CARDAMOM SEEDS.
Definition.—The *fruit* of Elettaria repens.
Natural Order.—Scitamineæ. *Habitat.*—East Indies.
Used in pulvis aromaticus.

Officinal Preparations.
Tinctura Cardamomi, ... f℥ ss–j.
Tinctura Cardamomi Composita, . . f℥ ss–j.

Officinal Name, PIPER. *English Name*, BLACK PEPPER.
Definition.—The *unripe fruit* of Piper nigrum.
Natural Order.—Piperaceæ. *Habitat.*—East Indies.
Contains an acrid resin, the alkaloid **piperine**, and a volatile oil.

Officinal Preparation.
Oleoresina Piperis, ... ♏ss–ij.

Officinal Name, CAPSICUM. *English Name*, RED, CAYENNE, AFRICAN, PEPPER.
Definition.—The *fruit* of Capsicum fastigiatum.

FOR THERAPEUTIC NOTES.

Natural Order.—Solanaceæ. *Habitat.*—West Indies.

Contains a resin, fixed and volatile oils.

May be used in **doses** of gr. ij-v, in pill.

Officinal Preparations.

Extractum Capsici Fluidum, ... ♏ij-x.
Oleoresina Capsici, ... gr. ss-j.
Tinctura Capsici, ... ♏v-xxx.
Emplastrum Capsici.

Officinal Name, ZINGIBER. *English Name,* GINGER.

Definition.—The *rhizome* of Zingiber officinale.

Natural Order.—Scitamineæ. *Habitat.*—East and West Indies.

Contains a volatile oil and an acrid resin.

Used in pulvis aromaticus and in pulvis rhei compositus.

Officinal Preparations.

Oleoresina Zingiberis, ... ♏ss-ij.
Extractum Zingiberis Fluidum, . . . ♏v-xv.
Tinctura Zingiberis, ... ʒ ss-j.
Syrupus Zingiberis, ... vehicle.
Trochisci Zingiberis, ... j = ♏ij of tincture.

Officinal Name, FŒNICULUM. *English Name,* FENNEL.

Definition.—The *fruit* of Fœniculum capillaceum.

Natural Order.—Umbelliferæ. *Habitat.*—South England.

ESSENTIALS OF MATERIA MEDICA. 251

Officinal Preparations.
Aqua Fœniculi, ... f℥j-f℥ss.
Oleum Fœniculi, ... ♏j-v.

Officinal Name, OLEUM CAJUPUTI. *English Name*,
OIL OF CAJUPUT.
Definition.—A volatile oil distilled from the *leaves* of Melaleuca leucadendron.
Natural Order.—Myrtaceæ. *Habitat.*—Molucca Islands.
Dose, ♏v-xv.
Used as parasiticide, externally.

Officinal Name, CARUM. *English Name*, CARAWAY.
Definition.—The *fruit* of Carum carvi.
Natural Order.—Umbelliferæ. *Habitat.*—Europe.
Used in compound tincture of cardamom.

Officinal Preparation.
Oleum Cari, ... ♏j-v.

Officinal Name, CORIANDRUM. *English Name*,
CORIANDER.
Definition.—The *fruit* of Coriandrum sativum.
Natural Order.—Umbelliferæ. *Habitat.*—Europe.
Dose, gr. v-xxx.

Officinal Preparation.
Oleum Coriandri, ... ♏j-v.

ESSENTIALS OF MATERIA MEDICA.   253

Officinal Name, ANISUM.   *English Name*, ANISE.

Definition.—The *fruit* of Pimpinella anisum.

Natural Order.—Umbelliferæ.   *Habitat.*— Europe.

Officinal Preparations.
Oleum Anisi, ... ♏ j–v.
Spiritus Anisi, ... ꝶ ss–ij.
Aqua Anisi, ... vehicle.

Officinal Name, OLEUM SASSAFRAS.   *English Name*, OIL OF SASSAFRAS.

Definition.—A volatile *oil* distilled from Sassafras.

Dose, ♏ ij–v, mostly as flavoring.

Officinal Name, AURANTII DULCIS CORTEX.   *English Name*, SWEET ORANGE PEEL.

Definition.—The *rind* of the *fruit* of Citrus aurantium.

Natural Order. — Rutaceæ.   *Habitat.* — Europe and United States.

Officinal Preparations.
Oleum Aurantii Corticis, ...
Spiritus Aurantii, ...
Spiritus Aurantii Compositus, . . .   } vehicles and flavors.
Tinctura Aurantii Dulcis, ...
Syrupus Aurantii, ...

Officinal Name, AURANTII AMARI CORTEX.   *English Name*, BITTER ORANGE PEEL.

Definition.—The *rind* of the *fruit* of Citrus vulgaris.

Natural Order.—Rutaceæ.

Officinal Preparations.
Extractum Aurantii Amari Fluidum, . f ʒ j.
Tinctura Aurantii Amari, ... f ʒ j–ij.

Officinal Name, AURANTII FLORES (Pharm., 1880).
English Name, ORANGE FLOWERS.

Definition.—The *flowers* of both the foregoing species.

Officinal Preparations.
Syrupus Aurantii Florum, ... vehicle.
Aqua Aurantii Florum Fortior, ... vehicle.
Oleum Aurantii Florum, ... flavor.

Officinal Name, LAVANDULA (Pharm., 1880). *English Name,* LAVENDER.

Definition.—The *fresh leaves* and *tops* of Lavandula officinalis.

Natural Order.—Labiatæ.

Officinal Preparations.
Oleum Lavandulæ Florum, ... ♏ j–v.
Tinctura Lavandulæ Composita, . . . f ʒ ss–ij.
Spiritus Lavandulæ, ... f ʒ ss–ij.

Officinal Name, SALVIA. *English Name,* SAGE.

Definition.—The *leaves* of Salvia officinalis.

Natural Order. —Labiatæ. *Habitat.* —United States.

Often used as a gargle in shape of an infusion.

ESSENTIALS OF MATERIA MEDICA. 257

Officinal Name, ROSMARINUS (Pharm., 1880). *English Name,* ROSEMARY.

Definition.—The *leaves* of Rosmarinus officinalis.

The oil is sometimes used in two to five drop doses.

Officinal Name, MENTHA PIPERITA. *English Name,* PEPPERMINT.

Definition.—The *leaves* and *tops* of Mentha piperita.

Officinal Name, MENTHA VIRIDIS. *English Name,* SPEARMINT.

Definition.—The *leaves* and *tops* of Mentha viridis.

Both Mentha piperita and Mentha viridis belong to *Natural order,* Labiatæ. *Habitat.*—United States.

The oil, water, and spirit of both peppermint and spearmint are officinal.

Officinal Preparations.

Oleum Menthæ Piperitæ, ...	
Oleum Menthæ Viridis, ...	vehicles.
Aqua Menthæ Piperitæ, ...	
Aqua Menthæ Viridis, ...	
Spiritus Menthæ Piperitæ, ... f ℥ ss.	
Spiritus Menthæ Viridis, ... f ℥ ss.	

Officinal Name, MELISSA. *English Name*, BALM.

Definition.—The *leaves* and *tops* of Melissa officinalis.

Natural Order.—Labiatæ.

Used in infusion.

Officinal Name, CALAMUS. *English Name*, SWEET FLAG.

Definition.—The *rhizome* of Acorus calamus.

Natural Order.—Aroideæ. *Habitat.*—United States.

Officinal Preparation.

Extractum Calami Fluidum, ... f ℨ ss–j.

AROMATIC BITTERS.

Officinal Name, ANTHEMIS. *English Name*, CHAMOMILE, OR CHAMOMILE FLOWERS.

Definition.—The *flower-heads* of Anthemis nobilis.

Natural Order.—Compositæ. *Habitat.*—Europe and America.

Mostly used as an infusion, the so-called chamomile tea (German Brust-Thee), breast-tea.

Dose, ℥ss of the flowers to Oj of water.

Officinal Name, SERPENTARIA. *English Name*, VIRGINIA SNAKEROOT.

Definition.—The *rhizome* and *rootlets* of Aristolochia serpentaria and of Aristolochia reticulata.

ESSENTIALS OF MATERIA MEDICA. 261

Natural Order.—Aristolochiaceæ. *Habitat.*—United States.
Used in tinctura cinchonæ composita.

Officinal Preparations.
Extractum Serpentariæ Fluidum, . . ♏x-f℥ss.
Tinctura Serpentariæ, ... f℥ss-ij.

Officinal Name, CASCARILLA. *English Name*, CASCARILLA.

Definition.—The *bark* of Croton eluteria.
Natural Order.—Euphorbiaceæ. *Habitat.*—West Indies.
Used mostly in infusion.

FAMILY II.—EMETICS.

Emetics include those drugs which cause emesis or vomiting.

VEGETABLE EMETICS.

Officinal Name, IPECACUANHA. *English Name*, IPECAC.

Definition.—The *root* of Cephaëlis ipecacuanha.
Natural Order.—Rubiaceæ. *Habitat.*—Brazil.
Its activity is due to the alkaloid **emetine**.
Emetic **dose** of ipecac in powder, gr. x–xxx; as diaphoretic and expectorant, gr. ⅛–j.

Officinal Preparations.
Syrupus Ipecacuanhæ (used almost entirely for children), emetic dose, f℥j–iv; diaphoretic and expectorant, . . ♏iij–xx.

ESSENTIALS OF MATERIA MEDICA. 263

Vinum Ipecacuanhæ, emetic dose, . . f ʒ j–ij.
Extractum Ipecacuanhæ Fluidum, emetic dose, ... ♏xx–xl.
Trochisci Ipecacuanhæ, ... 1=gr. ¼ of ipecac.
Tinctura Ipecacuanhæ et Opii, . . . ⎫
Pulvis Ipecacuanhæ et Opii, ... ⎬ See Opium.
Trochisci Morphinæ et Ipecacuanhæ, ⎭

Officinal Name, SANGUINARIA. *English Name*, BLOODROOT.

Definition.—The *rhizome* of Sanguinaria canadensis.

Natural Order.—Papaveraceæ. *Habitat.*—Widely diffused.

Contains the **alkaloids** chelerythrine (most abundant), homochelidonine, **sanguinarine**, and protopine.

Dose, as emetic, gr. x–xxx; rarely used in crude form.

Officinal Preparations.
Extractum Sanguinariæ Fluidum, . . . ♏j–v.
Tinctura Sanguinariæ, ... ♏xv–xxx.

Officinal Name, APOMORPHINÆ HYDROCHLORAS.
English Name, HYDROCHLORATE OF APOMORPHINE.

Definition.—The hydrochlorate of an artificial alkaloid prepared from morphine or codeine.

Dose, emetic,—generally given hypodermically, —gr. $\frac{1}{12}$–$\frac{1}{8}$; expectorant, gr. $\frac{1}{20}$–$\frac{1}{10}$.

Squill is occasionally used as an emetic.

FOR THERAPEUTIC NOTES.

Mustard Flour—the ordinary mustard of the grocer—is one of the emetics most frequently employed. It is given in warm water, about ʒij to Oj, repeated if necessary in two to five minutes.

MINERAL EMETICS.

Tartar Emetic,—very slow and depressing, but extremely persistent. Not much used.

Dose, gr. ss–ij.

Sulphate of zinc acts promptly, is purely mechanical, and produces no irritation. It acts well in combination with ipecac, say 30 grains of sulphate of zinc with 50 grains of ipecac, and then one-half of the above mixture every fifteen minutes to effect desired result.

Sulphate of copper is more irritating than sulphate of zinc, which is preferable in every way.

Dose, gr. v–x, not to be repeated.

Alum (powdered) has been used in ʒj dose in syrup or molasses for children (as in membranous croup), but is considered unreliable by Dr. Wood.

FAMILY III.—CATHARTICS.

Purgatives or cathartics are drugs producing purgation or catharsis by increasing intestinal secretion or peristaltic action.

They include (1) laxatives, (2) purges, (3) hydragogues, and (4) drastics.

Laxatives simply cause a mild evacuation of the

bowels, and do not produce purgation even when given in large doses.

Officinal Name, TAMARINDUS. *English Name*, TAMARIND.

Definition.—The preserved *pulp* of the *fruit* of Tamarindus indica.

Natural Order.—Leguminosæ. *Habitat.*—East and West Indies.

Contains citric, malic, and tartaric acids.

Dose, ʒss–j, as laxative.

Used in confection of senna.

Officinal Name, MANNA.

Definition.—The *concrete, saccharine exudation* of Fraxinus ornus.

Natural Order.—Oleaceæ. *Habitat.*—Sicily.

Contains *mannite*, an active crystalline principle.

Dose, for adult, ʒss–ij ; child, ʒj–ʒss.

Officinal Name, CASSIA FISTULA. *English Name*, PURGING CASSIA.

Definition.—The *fruit* of Cassia fistula.

Natural Order.—Leguminosæ. *Habitat.*—North America.

Dose, ʒj–ij.

Used in confection of senna.

Officinal Name, FRANGULA. *English Name*, BUCKTHORN.

Definition.—The *bark* of Rhamnus frangula, collected at least one year before being used.

Contains *franguline*, an active principle, and *emodine* (glucoside).

Officinal Preparation.
Extractum Frangulæ Fluidum, ... f℥ss–ij.

The *bark* of Rhamnus purshiana, or California buckthorn (**cascara sagrada**) is much oftener used than the above.

Officinal Preparation.
Extractum Rhamni Purshianæ Fluidum, ... ♏x–f℥j.

Officinal Name, EUONYMUS. *English Name,* WAHOO.

Definition.—The *bark* of the *root* of Euonymus atropurpureus.

Natural Order.—Celastrineæ. *Habitat.*—United States.

Contains *euonymin*,* a bitter principle.

Dose, gr. ij–iv.

Officinal Preparation.
Extractum Euonymi, ... gr. ij–vj.

Officinal Name,   *English Name,*
MAGNESIA.   LIGHT MAGNESIA.
MAGNESIA PONDEROSA.   HEAVY MAGNESIA.

Definition.—These differ only in physical characteristics.

Dose, gr. x–℥ij.

---

* Not officinal.

Officinal Preparations.
Magnesii Carbonas, ... gr. x–ʒij.
Magnesii Citras Effervescens, ... gr. x–ʒij.
Liquor Magnesii Citratis, ... fʒiv–viij.
Magnesii Sulphas (*Epsom salt*), more
   active than the above, ... ʒij–ʒj.

SULPHUR. BRIMSTONE.*

Dose, gr. x–xx, t. i. d.

Officinal Preparations.
Sulphur Sublimatum (flowers of sul-
phur), often given in molasses, . . . ʒj–iv.
Unguentum Sulphuris.
Sulphur Lotum (washed sulphur), . . ʒj–iv.
Sulphuris Iodidum.
Sulphur Præcipitatum (precipitated sul-
phur), ... ʒj–iv.

Calx sulphurata et potassa sulphurata—sulphurated lime and sulphurated potassium—are occasionally, though rarely, used.

Dose, gr. $\frac{1}{10}$–$\frac{1}{4}$.

Purges.—Medicines which cause brisk catharsis but are not poisonous even in large doses.

Officinal Name, OLEUM RICINI. *English Name,* CASTOR OIL.

Definition.—A cold-expressed oil from the *seeds* of Ricinus communis.

---
* Not officinal.

Natural Order.—Euphorbiaceæ. *Habitat.*—India.
The seeds contain *ricin*, an active poisonous principle which, however, is not communicated to the oil.

Dose, fʒij-f℥j.

Officinal Name, HYDRARGYRUM. *English Name*, MERCURY.

Blue mass and calomel are used as cathartics. (See Mercury.)

Officinal Name, RHEUM. *English Name*, RHUBARB.

Definition.—The *root* of Rheum officinale.
Natural Order.—Polygonaceæ. *Habitat.*—China.
Dose, in powder as stomachic, gr. j–v; cathartic, gr. xx–ʒj.

Officinal Preparations.

Extractum rhei,	gr. v–x.
Extractum rhei fluidum,	♏x–xxx.
Tinctura rhei,	⎫
Tinctura rhei aromatica,	⎬ fʒss–ij.
Tinctura rhei dulcis,	⎭
Mistura rhei et sodæ,	f℥ss–ij.
Syrupus rhei (for infants),	fʒj.
Syrupus rhei aromaticus (for infants),	fʒj.
Pilulæ rhei,	āā gr. iij.
Pilulæ rhei compositæ (rhubarb, gr. ij; aloes, gr. jss).	
Pulvis rhei compositus (ginger, 1; rhubarb, 2; magnesia, 6).	

ESSENTIALS OF MATERIA MEDICA. 275

Officinal Name, JUGLANS. *English Name*, BUTTERNUT.
Definition.—The *bark* of the *root* of Juglans cinerea, collected in the autumn.
Natural Order.—Juglandaceæ. *Habitat.*—United States.
Officinal Preparation.
Extractum Juglandis, ... gr. x–xxx.

Officinal Name, ALOE (Pharm., 1880), ALOE BARBADENSIS, ALOE SOCOTRINA. *English Name*, ALOES.
Definition.—The *inspissated juice* of the *leaves* of Aloe vera (Barbadoes aloes) and Aloe Perriyi (socotrine aloes).
Natural Order.—Liliaceæ. *Habitat.*—Socotra, Zanzibar.
Contains *aloinum* (*aloin*), a neutral principle.

Officinal Preparations.
Aloe Purificata, ... gr. ss–x.
Extractum Aloes, ... gr. ss–v.
Pilulæ Aloes, soap and aloes, . . āā gr. ij.
Pilulæ Aloes et Asafœtidæ, aloes, asafetida, and soap, of each, 1⅓ grs.
Pilulæ Aloes et Mastiches (*Lady Webster's pills*), aloes, 2 grs.; mastic and rose, of each, ½ gr.
Pilulæ Aloes et Ferri, aloes and ferrous sulphate, of each, 1 gr.
Pilulæ Aloes et Myrrhæ, aloes, 2 grs.; myrrh, 1 gr.; aromatic powder, ½ gr.
Tinctura Aloes, ... f ʒ ss–ij.
Tinctura Aloes et Myrrhæ, ... f ʒ ss–ij.
Aloinum, ... gr. 1⁄10–j.

Officinal Name, SENNA. *English Name*, SENNA.

Definition.—The *leaflets* of Cassia acutifolia and Cassia angustifolia.

Natural Order.—Leguminosæ. *Habitat.*—Egypt and Arabia.

Officinal Preparations.

Confectio Sennæ,	ℨj–iv.
Syrupus Sennæ,	fʒj–iv.
Extractum Sennæ Fluidum,	fʒj–iv.
Infusum Sennæ Compositum (black draught),	fʒij–f℥ss.

Used also in pulvis glycyrrhizæ compositus (compound liquorice powder).

Hydragogues "(including the **salines**) produce large watery stools without much irritation."

Officinal Preparations of Magnesia used as Purgatives.

Magnesii Sulphas (Epsom salt),	gr. x–ʒj.
Magnesii Citras Effervescens,	gr. x–ʒj.
Liquor Magnesii Citratis,	f℥ij–iv.

Officinal Name, SODII SULPHAS. *English Name*, GLAUBER'S SALT.

Definition.—Not much used; practically the same as sulphate of magnesia, but of more disagreeable taste.

Officinal Name, SODII PHOSPHAS. *English Name*, PHOSPHATE OF SODIUM.

Dose, gr. x–ʒiv.

Officinal Name, POTASSII ET SODII TARTRAS.
English Name, ROCHELLE SALT.

Definition.—Occurs in large crystals, but usually kept in powdered form.

Dose, ℨss–ij, in water.

Officinal Name, PULVIS EFFERVESCENS COMPOSITUS.
English Name, SEIDLITZ POWDER.

Definition.—Contains two papers: the white is made of 35 grains of tartaric acid, the blue of 40 grains of sodium bicarbonate and 120 grains of Rochelle salt.

Dose, one powder. Each paper to be dissolved in a separate glass of water, mix, and drink while effervescing.

Drastics.—Active, irritant vegetable cathartics; in sufficient amount may cause death.

Officinal Name, JALAPA. *English Name,* JALAP.

Definition.—The *tuberous root* of Ipomœa jalapa. *Natural Order.* — Convolvulaceæ. *Habitat.* — Mexico.

Contains an active resin, to which its properties are due.

Officinal Preparations.

Extractum Jalapæ, ... gr. iij–x.
Resina Jalapæ, ... gr. ij–v.
Pulvis Jalapæ Compositus, jalap, 35
   per cent.; potassium bitartrate, 65 per
   cent., ... gr. x–ʒj.

ESSENTIALS OF MATERIA MEDICA. 281

Officinal Name, COLOCYNTHIS. *English Name*,
COLOCYNTH.

Definition.—The *fruit* of Citrullus colocynthis, *deprived of its rind.*
Natural Order.—Cucurbitaceæ. *Habitat.*—Africa.

Officinal Preparations.

Extractum Colocynthidis, ... gr. ij–v.
Extractum Colocynthidis Compositum, laxative dose, gr. j–iij; purgative dose, ... gr. x–xx.
Pilulæ Catharticæ Compositæ, contain compound ext. of colocynth, 80 gm.; mild mercurous chloride, 60 gm.; ext. of jalap, 30 gm.; gamboge, 15 gm.; water, q. s., to make 1000 pills, ... j–iij pills.

Officinal Name, SCAMMONIUM. *English Name*,
SCAMMONY.

Definition.—A *resinous exudation* from the living *root* of Convolvulus scammonia.
Natural Order. — Convolvulaceæ. *Habitat.* — Syria.
Its activity is due to the resin *scammonin.*
Dose, gr. v–xx.

Officinal Preparation.

Resina Scammonii, ... gr. ij–x.

19

ESSENTIALS OF MATERIA MEDICA. 283

Officinal Name, PODOPHYLLUM. *English Name*, MAY APPLE.

Definition.—The *rhizome* and *roots* of Podophyllum peltatum.

Natural Order.—Berberideæ. *Habitat.*—United States.

Contains the alkaloid **berberine** and several resins, to which its activity is due.

Dose, in powder, gr. x–xx.

Officinal Preparations.

Extractum Podophylli, ... gr. v–x.
Extractum Podophylli Fluidum, . . . ℥ x–xx.
Resina Podophylli (podophyllin), . . gr. ⅛–¼.

Officinal Name, ELATERINUM. *English Name*, ELATERIN.

Definition.—A neutral principle obtained from *elaterium*, a substance deposited by the *juice* of Ecballium elaterium (squirting cucumber).

Natural Order.—Cucurbitaceæ. *Habitat.*—Europe.

Dose, gr. 1/12–1/6.

Officinal Preparation.

Trituratio Elaterini, ... gr. ¼–j.

Officinal Name, CAMBOGIA. *English Name*, GAMBOGE.

Definition.—A *gum resin* obtained from Garcinia Hanburii.

Natural Order.—Guttiferæ. *Habitat.*—Siam.
Used in pilulæ catharticæ compositæ.

Officinal Name, OLEUM TIGLII. *English Name,*
CROTON OIL.
Definition.—A *fixed oil* expressed from the *seed*
of Croton tiglium.
Natural Order.—Euphorbiaceæ. *Habitat.*—India.
Dose, ♏j–ij.
Used externally as counter-irritant.

FAMILY IV.—DIURETICS.

Medicines which increase the flow of urine. They include (1) the hydragogue diuretics, which simply increase the flow of water from the kidneys, and are therefore useful in dropsy. (2) Refrigerant diuretics, which exert a marked sedative action, and so modify the secretion that they render the urine less irritant. (3) Alterative diuretics, whose active principles are eliminated by the kidneys and thus act on the mucous surfaces over which they pass.

HYDRAGOGUE DIURETICS.

Officinal Name, SCILLA. *English Name,* SQUILL.
Definition.—The sliced *bulb* of Urginea maritima.
Natural Order. — Liliaceæ. *Habitat.* — South Europe.
Dose, in powder, gr. j–ij.

Officinal Preparations.

Acetum Scillæ, ...	♏x–f℥ss.
Extractum Scillæ Fluidum, ...	♏j–ij.
Tinctura Scillæ, ...	♏x–f℥ss.
Syrupus Scillæ, ...	f℥ss–j.

Syrupus Scillæ Compositus (*Cox's hive syrup*), 2 parts in 1000 of tartar emetic, ♏v–xxx.

Officinal Name, SCOPARIUS.   *English Name*, BROOM.

Definition.—The *tops* of Cytisus scoparius. *Natural Order.*—Leguminosæ. *Habitat.*—Europe. Contains the alkaloid **sparteine** and scoparin.

Officinal Preparation.

Extractum Scoparii Fluidum, ... ♏ij–x.

Calomel in large doses is occasionally used as a diuretic.

The alkaloid **theobromine,*** used as the salicylate or double salicylate of sodium and theobromine (diuretine), in daily doses of gr. xv–c, is employed at times, and is a valuable diuretic.

BLATTA.*

Definition.—The dried bodies of Blatta orientalis (cockroach) have often been used as a popular remedy in dropsy.

Dose, gr. xv–xxx in twenty-four hours.

---

* Not officinal.

ESSENTIALS OF MATERIA MEDICA. 289

Officinal Name, SPIRITUS ÆTHERIS NITROSI. *English Name*, SPIRIT OF NITROUS ETHER.

Definition.—An alcoholic solution of ethyl nitrite, yielding not less than 11 times its own volume of nitrogen dioxid.

Dose, as diuretic, fʒj–iv.

Caffeine, jaborandi, strophanthus, and digitalis are all valuable diuretics (see Doses, etc., under proper headings).

REFRIGERANT DIURETICS.

POTASSIUM.*

Definition.—Obtained from vegetable *ash*, *nitre*, and from the *argol* or tartar deposited by wine.

Officinal Preparations.

Potassa (caustic).
Liquor Potassæ (five per cent. potassium
  hydrate), ... ♏v–xx.
Liquor Potassii Citratis, citric acid, six
  parts ; potass. bicarb., eight parts in
  100, ... fʒss.
Potassii Acetas, ... ⎫
Potassii Chloras, ... ⎪
Potassii Citras, ... ⎪
Potassii Carbonas, ... ⎬ gr. v–xxx.
Potassii Bicarbonas, ... ⎪
Potassii Nitras (saltpetre), . . . ⎪
Potassii Bitartras (*cream of tartar*), ⎭

---

* Not officinal.

Potassii et Sodii Tartras (Rochelle salt), ℥j.
Potassii Sulphas, ... ℥j.
Trochisci Potassii Chloratis, ... 1 = gr. v.
Liquor Potassii Citratis (Citric acid
 neutralized by potass. bicarb.), . . f℥ss-j.
Charta Potassii Nitratis, made from the nitrate (nitre).
Potassii Bichromas.
Potassii Ferrocyanidum.
Potassa Sulphurata.
Potassa cum Calce (caustic).

Officinal Name, LITHII CARBONAS. *English Name*,
LITHIUM CARBONATE.

Dose, gr. v-xv, t. i. d.

Besides the carbonate, which is diuretic, there are officinal:

Lithii Salicylas, ...
Lithii Bromidum, ... } gr. v-xx.
Lithii Benzoas, ...
Lithii Citras, ...

PIPERAZINE.*

Definition.—Occurs in small, glossy crystals. Its value depends entirely on its solvent power over uric acid.

May be given hypodermically in two per cent. solution, or gr. xv, in twenty-four hours, in water, as it is highly hygroscopic.

---

* Not officinal.

FOR THERAPEUTIC NOTES.

ESSENTIALS OF MATERIA MEDICA. 293

Strontium iodide, lactate, and bromide, the salts of strontium, are sometimes employed.
Dose, gr. xxx in twenty-four hours.

ALTERATIVE DIURETICS.

Officinal Name, BUCHU. *English Name,* BUCHU.
Definition.—The *leaves* of Barosma betulina and Barosma crenulata.
Natural Order.—Rutaceæ. *Habitat.*—Africa.

Officinal Preparation.
Extractum Buchu Fluidum, ... ♏xx–f℥j.

Officinal Name, PAREIRA. *English Name,* PAREIRA BRAVA.
Definition.—The *root* of Chondodendron tomentosum.
Natural Order.—Menispermaceæ. *Habitat.*—Brazil.

Officinal Preparation.
Extractum Pareiræ Fluidum, ... f℥ss–ij.

Officinal Name, UVA URSI. *English Name,* BEARBERRY.
Definition.—The *leaves* of Arctostaphylos uva-ursi.
Natural Order.—Ericaceæ. *Habitat.*—United States and Europe.

Contains the glucoside *arbutin,* to which its activity is due.

Officinal Preparations.

Extractum Uvæ Ursi, ... gr. x-xv.
Extractum Uvæ Ursi Fluidum, ... f ℥ ss–ij.

Officinal Name, CHIMAPHILA. *English Name*, PIPSISEWA.

Definition.—The *leaves* of Chimaphila umbellata. *Natural Order.*—Ericaceæ. *Habitat.*—United States.

Officinal Preparation.

Extractum Chimaphilæ Fluidum, . . f ℥ ss–j.

Officinal Name, JUNIPERUS (Pharm., 1880). *English Name*, JUNIPER.

Definition.—The *fruit* of Juniperus communis. *Natural Order.*—Coniferæ. *Habitat.*—United States and Europe.
Contains a volatile oil.

Officinal Preparations.

Oleum Juniperi, ... ♏ij–v.
Spiritus Juniperi (five per cent. of oil), f ℥ ss–ij.
Spiritus Juniperi Compositus (contains oils of juniper, caraway, and fennel), f ℥ j–ij.

Officinal Name, OLEUM ERIGERONTIS. *English Name*, OIL OF ERIGERON.

Definition.—A volatile oil distilled from the *fresh flowering herb* of Erigeron canadense or Canada fleabane.

Natural Order.—Compositæ.
Dose, ♏v–f℥ss.

Officinal Name, OLEUM SANTALI. *English Name*, OIL OF SANDALWOOD.
Definition.—A volatile oil distilled from the *wood* of Santalum album.
Natural Order.—Santalaceæ. *Habitat.*—Asia and Australia.
Dose, ♏ij–x.

Officinal Name, ZEA. *English Name*, CORN-SILK.
Definition.—The *styles* and *stigmas* of Zea mays.
Natural Order. — Gramineæ. *Habitat.* — Indigenous.

Preparation.
Extractum Zeæ Fluidum, ... f℥ss–j.

Officinal Name, TEREBINTHINA. *English Name*, TURPENTINE.
Definition.—A concrete *oleoresin* from the Pinus palustris.
Natural Order.—Coniferæ. *Habitat.*—Indigenous.

Officinal Preparations.
Oleum Terebinthinæ (wrongly called
 spirit of turpentine), ... ♏v–xv.
Linimentum Terebinthinæ.

Officinal Name, TEREBINTHINA CANADENSIS. *English Name,* CANADA TURPENTINE, CANADA BALSAM, BALSAM OF FIR.

Definition.—A liquid *oleoresin* from Abies balsamea.

Rarely used.

Officinal Name, COPAIBA. *English Name,* COPAIBA.

Definition.—The *oleoresin* of Copaiba Langsdorffii.

Natural Order.—Leguminosæ. *Habitat.*—South America.

Dose, ♏xx, t. i. d.

Officinal Preparations.
Massa Copaibæ, ... ♏v-xx.
Oleum Copaibæ, ... gr. v-xv.
Resina Copaibæ, ... gr. v-xv.

Officinal Name, CUBEBA. *English Name,* CUBEBS.

Definition.—The unripe *fruit* of Piper cubeba.

Natural Order.—Piperaceæ. *Habitat.*—East Indies.

Contains cubebic acid, a volatile oil, and the neutral principle *cubebin.*

Dose, in powder, ʒss-j.

Officinal Preparations.
Oleum Cubebæ, ... ♏v-xv.
Tinctura Cubebæ, ... f℥ss-ij.
Extractum Cubebæ Fluidum, ... ♏x-xxx.
Oleoresina Cubebæ, ... ♏v-xv.

ESSENTIALS OF MATERIA MEDICA. 301

Officinal Name, MATICO. *English Name*, MATICO.
Definition.—The *leaves* of Piper angustifolium.
Natural Order.—Piperaceæ. *Habitat.*—Peru.
Contains a volatile oil, resin, and bitter principle.

Officinal Preparations.
Extractum Matico Fluidum, ... f ℥ ss–j.
Tinctura Matico, ... f ℥ ss–ij.

FAMILY V.—DIAPHORETICS.

Medicines which increase the flow of perspiration, acting on the skin directly or through the system; they include: (1) Nauseating diaphoretics, (2) refrigerant diaphoretics, and (3) simple diaphoretics.

The **nauseating diaphoretics** include tartar emetic, ipecac and its preparations, notably Dover's powder.

The **refrigerant diaphoretics** include aconite, veratrum viride, the cardiac depressants, and, best of all, potassium citras—either as neutral mixture or effervescing draught.

The **simple diaphoretics** include:

Officinal Name, PILOCARPUS. *English Name*, JABORANDI.

Definition.—The *leaflets* of Pilocarpus selloanus and Pilocarpus jaborandi.
Natural Order.—Rutaceæ. *Habitat.*—Brazil.
Contains the alkaloid, **pilocarpine**.
Dose, of crude drug, gr. v to xl.

Officinal Preparations.
Extractum Pilocarpi Fluidum, ... ♏v-l.
Pilocarpinæ Hydrochloras, ... gr. ⅛-¼.

Officinal Name, LIQUOR AMMONII ACETATIS. *English Name*, SPIRIT OF MINDERERUS.
Definition.—Dilute acetic acid, neutralized by carbonate of ammonium. Valuable as a basis for fever mixtures.
Dose, f ʒj-f ʒss.

Officinal Name, SPIRITUS ÆTHERIS NITROSI. *English Name*, SPIRIT OF NITROUS ETHER.
Definition.—An alcoholic solution of ethyl nitrite yielding not less than 11 times its own volume of nitrogen dioxid.
Dose, as diaphoretic, f ʒj-f ʒss.

FAMILY VI.—EXPECTORANTS.
Remedies which cause an increase or modification in the amount of secretion from the larger tubes of the respiratory tract, and facilitate the expulsion thereof.
The nauseating expectorants are lobelia, tartar emetic, and ipecac. (See doses elsewhere.)

Officinal Name, GRINDELIA. *English Name*, GRINDELIA.
Definition.—The *leaves* and *flowering tops* of Grindelia robusta and Grindelia squarrosa.

ESSENTIALS OF MATERIA MEDICA. 305

Natural Order.—Compositæ. *Habitat.*—United States.
Officinal Preparation.
Extractum Grindeliæ Fluidum, . . . ℳx–f℈ij.

STIMULATING EXPECTORANTS.

Ammonium chloride (ammonii chloridum), in **doses** of gr. v–xx, is a valuable expectorant.

Officinal Name, SENEGA. *English Name*, SENEGA.
Definition.—The *root* of Polygala senega.
Natural Order.—Polygaleæ. *Habitat.*—United States.
Contains the glucoside *senegin*, and polygallic acid.
Officinal Preparations.
Extractum Senegæ Fluidum, ... ℳx–xxx.
Syrupus Senegæ, ... f℈ss–ij.

Officinal Name, AMMONIACUM. *English Name*, AMMONIAC.
Definition.—A *gum resin* from Dorema ammoniacum.
Natural Order.—Umbelliferæ. *Habitat.*—Persia.
Dose, gr. x–xx.
Officinal Preparations.
Emplastrum Ammoniaci cum Hydrargyro.
Emulsum Ammoniaci, ... f℥ss–j.

SULPHURETTED HYDROGEN.*
Rarely used.

Officinal Name, BENZOINUM. *English Name,* BENZOIN.
Definition.—A *balsamic resin* obtained from Styrax benzoin.
Natural Order.—Styraceæ. *Habitat.*—Peru.
Contains *benzoic acid,* a volatile oil, and a resin.

Officinal Preparations.

Adeps Benzoinatus,	external use.
Tinctura Benzoini,	f℥ss-ij.
Tinctura Benzoini Composita,	f℥ss-ij.
Acidum Benzoicum,	gr. x-℥ss.
Ammonii, Lithii, Sodii, } Benzoas,	gr. x-xv.

Officinal Name, BALSAMUM PERUVIANUM. *English Name,* BALSAM OF PERU.
Definition.—A *balsam* from Toluifera pereiræ.
Natural Order.—Leguminosæ. *Habitat.*—South America.
Dose, ℨss.

Officinal Name, BALSAMUM TOLUTANUM. *English Name,* BALSAM OF TOLU.
Definition.—A *balsam* obtained from Toluifera balsamum.

---

* Not officinal.

Natural Order.—Leguminosæ. *Habitat.*—Central America.

Officinal Preparations.

Tinctura Tolutana, ... f ℨ j–iij.
Syrupus Tolutanus, ... f ℨ j–ij.

Also used in the compound tincture of benzoin. Mainly used as vehicles.

Officinal Name, MYRRHA. *English Name*, MYRRH.
Definition.—A *gum-resin* from Commiphora myrrha.
Natural Order.—Burseraceæ. *Habitat.*—Arabia, Africa.

Officinal Preparations.

Pilulæ Aloes et Myrrhæ, ... j–iij pills.
Tinctura Aloes et Myrrhæ, ... ⎫
Tinctura Myrrhæ, ... ⎬ f ℨ ss–ij.
Mistura Ferri Composita, ... f ℨ ss.

Officinal Name, ALLIUM. *English Name*, GARLIC.
Definition.—The *bulb* of Allium sativum.
Natural Order.—Liliaceæ. *Habitat.*—Indigenous.

Officinal Preparation.

Syrupus Allii, ... f ℨ j–ij.

Syrup of squill and **compound syrup of squill** are also sometimes used as stimulating expectorants in **doses** of f℥ss.

Officinal Name, PIX LIQUIDA. *English Name*, TAR.

Definition.—The empyreumatic *oleoresin* obtained by destructive distillation of the *wood* of Pinus palustris and other species of Pinus.

Natural Order.—Coniferæ. *Habitat.*—Indigenous.

On distillation it yields pitch, oil of tar, and pyroligneous acid.

Officinal Preparations.

Oleum Picis Liquidæ, ... external use.
Syrupus Picis Liquidæ, ... f℥j-iv.
Unguentum Picis Liquidæ, 50 per cent. tar.

Officinal Name, TEREBENUM. *English Name*, TEREBENE.

Definition.—A clear, colorless *fluid*, obtained by the action of sulphuric acid on oil of turpentine, and then distilled.

Consists mostly of pinene and very small amounts of terpinene and dipentene.

Dose, ♏x, in capsule or emulsion, t. i. d.

FAMILY VII.—EMMENAGOGUES.

Remedies used to increase or re-establish the menstrual flow.

They include tonic and stimulating emmenagogues.

The tonic emmenagogues are *iron*, *myrrh*, and *aloes*. Iron and myrrh act simply by their general

ESSENTIALS OF MATERIA MEDICA. 313

tonic action and are largely used in anemic affections of the menstrual flow. The effect of aloes is due solely to stimulation of the rectum.

STIMULATING EMMENAGOGUES.

Officinal Name, SABINA. *English Name*, SAVINE.

Definition.—The *tops* of Juniperus sabina.
Natural Order.—Coniferæ. *Habitat.*—Europe.
Contains a *volatile oil*, to which its activity is due.

Officinal Preparations.

Oleum Sabinæ, ... ♏v–x.
Extractum Sabinæ Fluidum, ... ♏v–xv.

RUTA. RUE.*

Definition.—The *leaves* of Ruta graveolens.
Natural Order.—Rutaceæ. *Habitat.*—Europe.
Contains a *volatile oil.*

Preparation.

Oleum Rutæ, ... ♏ij–vj.

APIOL.*

Definition.—A liquid, neutral principle from Petroselinum sativum—ordinary parsley.
Natural Order.—Umbelliferæ.
Dose, ♏iij–x, in capsule.

Potassii permanganas, in doses of gr. j–ij, t. i. d. ; **cantharides,** in the form of the tincture,

---
* Not officinal.

♏iij-v, and the **ammoniated tincture of guaiac**, in doses of f℥ss-j, have often been employed with success.

Officinal Name, TANACETUM. *English Name*, TANSY.
Definition.—The *leaves* and *tops* of Tanacetum vulgare.
Natural Order.—Compositæ. *Habitat.*—Indigenous.
Contains a *volatile oil* and bitter principle.
Dose, gr. x-xx. The oil is dangerous, dose ♏j-v.

Officinal Name, HEDEOMA. *English Name*, PENNYROYAL.
Definition.—The *leaves* and *tops* of Hedeoma pulegioides.
Natural Order.— Labiatæ. *Habitat.*— United States.
Owes its activity to a *volatile oil*.

Officinal Preparation.
Oleum Hedeomæ, ... ♏j-v.

FAMILY VIII.—OXYTOCICS.

Remedies which increase uterine muscular contraction.

Officinal Name, ERGOTA. *English Name*, ERGOT.
Definition.—The *sclerotium* of Claviceps purpurea (*natural order*, Fungi), replacing the grain of

rye (Secale cereale, *natural order*, Gramineæ), a fungous growth from the diseased ovary of the rye. Should not be over one year old.

Dose, in powder, gr. xxx.

Officinal Preparations.
Extractum Ergotæ Fluidum, ... f ℨ ss–j.
Vinum Ergotæ, ... f ℨ ij–f ℥ ss.

Ergotin, a concentrated extract of ergot, is sometimes used ; gr. v equal to f ʒj of fluid extract.

Hydrastis and **hydrastinine hydrochlorate** have lately been used as oxytocics with good results. (See doses elsewhere.)

Officinal Name, GOSSYPII RADICIS CORTEX. *English Name*, COTTON ROOT BARK.

Definition.—The *bark* of the *root* of Gossypium herbaceum.

Natural Order.—Malvaceæ. *Habitat.*—United States.

Used in decoction (℥iv to Oj of water) f℥ij, repeated as needed.

Officinal Preparation.
Extractum Gossypii Radicis Fluidum, f ℨ ss- ij.

USTILAGO. SMUT OF INDIAN CORN.*

Definition.—Ustilago maydis (*natural order*,

---

* Not officinal.

ESSENTIALS OF MATERIA MEDICA. 319

Fungi), corn smut or corn ergot, a fungous growth on Zea mays (Indian corn).
Natural Order.—Gramineæ.
Dose, gr. xv–ʒj.

FAMILY IX.—SIALAGOGUES.

Those remedies which increase the flow of saliva and oral mucus.

Officinal Name, PYRETHRUM. *English Name*, PELLITORY.

Definition.—The *root* of Anacyclus pyrethrum.
Natural Order.—Compositæ. *Habitat.*—Europe.
Dose, ʒss–j to be chewed; or Tinctura Pyrethri, fʒss–ij.

FAMILY X.—ERRHINES.

Remedies acting on the nasal mucous membrane.

FAMILY XI.—EPISPASTICS.

Drugs used to produce blisters.

Officinal Name, CANTHARIS. *English Name*, CANTHARIDES, SPANISH FLIES.

Definition.—A beetle, Cantharis vesicatoria, inhabiting Spain, Italy, and Southern Europe.
Order.—Coleoptera.

The dried bodies contain a volatile oil and a neutral principle, *cantharidin*, to which is due the vesicating property.

ESSENTIALS OF MATERIA MEDICA. 321

Officinal Preparations.
Ceratum Cantharidis, ... external use.
Tinctura Cantharidis, ... ♏ij–v.
Collodium Cantharidatum, for blistering.
Emplastrum Picis Cantharidatum, warming plaster.

FAMILY XII.—RUBEFACIENTS.

Remedies causing powerful irritation and congestion of the skin surface, which is, however, of short duration.

Officinal Name,   *English Name,*
SINAPIS ALBA.   WHITE MUSTARD.
SINAPIS NIGRA.   BLACK MUSTARD.

Definition.—The *seed* of Brassica alba (white), and Brassica nigra (black),—mustard.
Natural Order.—Cruciferæ.   *Habitat.*—Europe.

Officinal Preparations (from black mustard).
Charta Sinapis, ... 6 grs. to square in.
Oleum Sinapis, ... ♏ij–v, diluted.
Linimentum Sinapis Compositum.

Capsicum, oleum terebinthinæ, and **ammonia** are also employed as rubefacients.

Officinal Name, PIX BURGUNDICA. *English Name,*
BURGUNDY PITCH.

Definition.—The prepared *resinous exudation* of Abies excelsa, or Norway spruce.
Natural Order.—Coniferæ.
Contains a resin and a volatile oil.

Officinal Preparations.
Emplastrum Picis Burgundicæ.
Emplastrum Picis Cantharidatum, warming plaster.

FAMILY XIII.—ESCHAROTICS.

Those remedies which, by contact, destroy either healthy or diseased tissues.

Officinal Name, ACIDUM CHROMICUM. *English Name*, CHROMIC ACID.

Definition.—Acicular crystals of deep-red color, very deliquescent, made by the action of sulphuric acid on potassium bichromate.

Apply with glass rod or platinum wire.

Officinal Name, BROMUM. *English Name*, BROMINE.

Definition.—A heavy, dark-red, mobile liquid ; very powerful caustic.

Besides the above are used :

> Alumen Exsiccatum.
> Acidum Sulphuricum.
> " Nitricum.
> " Hydrochloricum.
> " Arsenosum.
> Hydrargyri Chloridum Corrosivum.
> Cupri Sulphas.
> Zinci Sulphas.
> Zinci Chloridum.
> Potassa (caustic potash).
> Argenti Nitras Fusus (lunar caustic).
> Potassa cum Calce (Vienna paste).
> Liquor Potassæ, etc.

ESSENTIALS OF MATERIA MEDICA. 325

FAMILY XIV.—DEMULCENTS.

Bland substances, capable of soothing inflamed surfaces ; mostly of a gummy or mucilaginous consistency when mixed with water.

Officinal Name, ACACIA. *English Name*, GUM ARABIC.

Definition.—A gummy exudation from Acacia senegal.
Natural Order.—Leguminosæ. *Habitat.*—Africa and Australia.

Officinal Preparations.

Mucilago Acaciæ, }
Syrupus Acaciæ, } . . . . vehicles.

Officinal Name, TRAGACANTHA. *English Name*, TRAGACANTH.

Definition.—A *gummy exudation* from Astragalus gummifer.
Natural Order.—Leguminosæ. *Habitat.*—Asia Minor.

Officinal Preparation.

Mucilago Tragacanthæ, ... vehicle.

Officinal Name, ULMUS. *English Name*, SLIPPERY ELM.

Definition.—The *inner bark* of Ulmus fulva.
Natural Order.—Urticaceæ. *Habitat.*—United States.

Officinal Preparation.

Mucilago Ulmi, ... vehicle.

ESSENTIALS OF MATERIA MEDICA. 327

Officinal Name, CETRARIA. *English Name,* ICELAND MOSS.

Definition.—A *lichen,* Cetraria islandica, found in Iceland.
Natural Order.—Lichenes.

Officinal Preparation.

Decoctum Cetrariæ, ... f ℥ ss–iv.

Officinal Name, CHONDRUS. *English Names,* IRISH MOSS, CARRAGHEEN.

Definition.—*Fronds* of Chondrus crispus and Gigartina mamillosa.
Natural Order.—Algæ. *Habitat.*—New England and Ireland.
Nutrient and demulcent.

Officinal Name, GLYCYRRHIZA. *English Name,* LIQUORICE ROOT.

Definition.—The *root* of Glycyrrhiza glabra.
Natural Order.—Leguminosæ. *Habitat.*—Europe.
Contains the glucoside *glycyrrhizin.*

Officinal Preparations.

Extractum Glycyrrhizæ, ...
Extractum Glycyrrhizæ Fluidum, . . } for flavoring and as vehicles.
Extractum Glycyrrhizæ Purum, . .
Glycyrrhizinum Ammoniatum, ... gr. v–x.
Pulvis Glycyrrhizæ Compositus (cathartic),

contains senna, fennel, and washed
sulphur, ... ℨss–ij.
Mistura Glycyrrhizæ Composita (Brown
mixture) contains wine of antimony,
paregoric, and sweet spirit of nitre, . f℥ss–j.
Trochisci Glycyrrhizæ et Opii, 1 = ext.
of liquorice, 2 grs., ext. of opium,
gr. $\frac{1}{20}$.

Officinal Name, LINUM. *English Name*, FLAXSEED.
Definition.—The *seed* of Linum usitatissimum.
Natural Order.—Lineæ. *Habitat.*—Widely diffused.
Used as tea.
Officinal Preparation.
Oleum Lini (*flaxseed or linseed oil*).

Officinal Name, SASSAFRAS MEDULLA. *English Name*, SASSAFRAS PITH.
Definition.—The *pith* of Sassafras variifolium.
Natural Order.—Laurineæ. *Habitat.*—Europe and United States.
It yields a mucilage much used in the treatment of eye affections.

Officinal Name, ALTHÆA. *English Name*, MARSHMALLOW.
Definition.—The *root* of Althæa officinalis.
Natural Order.—Malvaceæ. *Habitat.*—United States.
Officinal Preparation.
Syrupus Althææ.

TAPIOCA.*

Definition.—The *fecula* obtained from the *root* of Janipha manihot.

Habitat.—South America.

Maranta (arrowroot), **sago**, and **hordeum** (barley), used mostly as food, are occasionally employed as demulcents.

FAMILY XV.—EMOLLIENTS.

Bland, fatty substances, which soothe and soften the skin.

Officinal Name, GLYCERINUM. *English Name*, GLYCERIN.

Definition.—A clear, colorless liquid of thick syrupy consistence; odorless; very sweet, and slightly warm to the taste. Obtained by the decomposition of vegetable or animal fats or fixed oils, and containing at least 95 per cent. of absolute glycerin.

Officinal Preparation.

Suppositoria Glycerini.

LANOLIN.*

Definition.—Purified fat of sheep's wool. Often used as an ointment base; more readily absorbed through the skin than most other fats, according to some authorities.

---

* Not officinal.

Officinal Name, ADEPS.   *English Name*, LARD.

Definition.—The prepared fat of Sus scrofa (hog), contains olein and stearin.

Officinal Preparations.
Adeps Benzoinatus.
Unguentum, lard, four parts ; yellow wax, one part.
Ceratum, lard, 70 per cent. ; white wax, 30 per cent.

Officinal Name, CETACEUM. *English Name*, SPERMACETI.

Definition.—A concrete fatty substance, obtained from Physeter macrocephalus (whale).

Officinal Preparations.
Ceratum Cetacei.
Unguentum Aquæ Rosæ.

Officinal Name,	*English Name*,
CERA FLAVA.	YELLOW WAX.
CERA ALBA.	WHITE WAX.

Definition.—Beeswax ; prepared by Apis mellifica (honey-bee).

Officinal Name, OLEUM THEOBROMATIS. *English Name*, BUTTER OF CACAO.

Definition.—A fixed *oil* expressed from the *seed* of Theobroma cacao. Used for suppositories and in ointments.

Natural Order.—Sterculiaceæ. *Habitat.*—South America.

Officinal Name,   *English Name,*
PETROLATUM MOLLE.   SOFT PETROLATUM.
PETROLATUM SPISSUM.   HARD PETROLATUM.

Definition.—A mixture of hydrocarbons obtained by the distillation of petroleum. When "petrolatum" is prescribed, it means always the soft variety.

FAMILY XVI.—DILUENTS.

These are substances (water and medicated waters) which are to be absorbed during their passage through the body, and so dilute its various fluids and excretions.

FAMILY XVII.—PROTECTIVES.

External applications to exclude air and protect inflamed surfaces.

Officinal Name, COLLODIUM.   *English Name,* COLLODION.

Definition.—A solution of pyroxylin or guncotton in alcohol and ether. The alcohol and ether evaporate rapidly and leave a translucent, flexible, adherent film on the skin which is impervious to air and water.

Officinal Preparations.

Collodium Flexile, five per cent. Canada turpentine,
    three per cent. castor oil, and 92 per cent. collodion.
Collodium Stypticum, 20 per cent. tannic acid.
Collodium Cantharidatum, blistering.

LIQUOR GUTTA-PERCHA (TRAUMATICINE).*
Contains nine per cent. of gutta-percha in commercial chloroform. Leaves a film of gutta-percha at place of application after evaporation of the chloroform, and in this manner various remedies are occasionally employed, especially in the treatment of skin diseases.

DIVISION II.
EXTRANEOUS REMEDIES.
FAMILY I.—ANTACIDS.
Remedies used to overcome excessive acidity.

SODIUM (Metal).*
Officinal Preparations.
Soda (caustic).
Liquor Sodæ, five per cent. sodium hydrate, ... ♏ij–x.
Sodii Bicarbonas, ... gr. v–xxx.
Sodii Carbonas, ... gr. v–xxx.
Trochisci Sodii Bicarbonatis, . . .
Sodii Acetas, ...
Sodii Benzoas, ...
Sodii Boras (borax), ...
Sodii Bromidum, ...
Sodii Chloras, ...
Sodii Chloridum (salt), ... gr. v–xxx.
Sodii Iodidum, ...
Sodii Phosphas, ...
Sodii Salicylas, ...
Sodii Sulphi Carbolas, ...
Sodii Sulphas (Glauber's salt), . . .

The liquor, carbonate, and bicarbonate are oftenest used.

---
* Not officinal.

CALCIUM (Metal).*

Officinal Preparations.

Calcii Bromidum, ...	
Calcii Chloridum, ...	gr. x–xxx.
Calcii Phosphas Præcipitatus, . . .	
Calcii Hypophosphis, ... gr. iij–v.	

Calcii Carbonas Præcipitatus.
Calx (quick-lime), caustic.
Liquor Calcis (lime-water).
Linimentum Calcis (*carron oil*), equal parts of lime-water and olive oil.

Calx Sulphurata, ... gr. $\frac{1}{10}$–ss.
Creta Præparata (prepared chalk), . . gr. x–ʒj.
Mistura Cretæ, ... f℥ss.
Syrupus Calcii Lactophosphatis, . . . f℥ss.
Syrupus Calcis, ... ♏x–fʒj.
Syrupus Hypophosphitum, ... fʒij–iv.
Syrupus Hypophosphitum cum Ferro, . fʒij–iv.

FAMILY II.—ANTHELMINTICS.

Remedies causing the expulsion or death of intestinal worms.

Officinal Name, SPIGELIA. *English Name*, PINKROOT.

Definition.—The *rhizome* and *rootlets* of Spigelia marilandica, or Carolina pink.

Natural Order.—Loganiaceæ. *Habitat.*—United States.

Officinal Preparation.
Extractum Spigeliæ Fluidum, . . . fʒss–j.

---

* Not officinal.

AZEDARACH.*

Definition.—The *bark* of the *root* of Melia azedarach (pride of China).
Natural Order.—Meliaceæ.
Used in f℥ss doses of decoction, ℥iv to Oj of water boiled to Oij.

Officinal Name, CHENOPODIUM. *English Name*, AMERICAN WORMSEED.

Definition.—The *fruit* of Chenopodium ambrosioides; variety, anthelminticum.
Natural Order.—Chenopodiaceæ. Its effects are due to the volatile oil it contains.
Dose, of oil, ♏v-x, for three-year-old child.

Officinal Name, CUSSO (BRAYERA, Pharm., 1880). *English Name*, KOUSSO.

Definition.—The female *inflorescence* of Hagenia abyssinica.
Natural Order.—Rosaceæ. *Habitat.*—Abyssinia.

Officinal Preparation.
Extractum Cusso Fluidum, ... f℥j–iij.

Officinal Name, SANTONICA. *English Name*, LEVANT WORMSEED.

Definition.—The unexpanded *flower-heads* of Artemisia pauciflora.

---
* Not officinal.

Natural Order.—Compositæ.   *Habitat.*—Europe and Asia.

Contains the neutral principle *santoninum* (*santonin*), to which its activity is due. Used for roundworms almost exclusively.

Dose, gr. ss–v for adult, and gr. ¼–ss for child.

Officinal Preparations.

Trochisci Santonini, ... I = gr. ss.
Sodii Santoninas (Pharm., 1880), . . gr. ij–v.

The sodium santoninate is more soluble, and therefore more dangerous than santonin.

Officinal Name, ASPIDIUM. *English Name*, MALE FERN —FILIX-MAS.

Definition.—The *rhizome* of Dryopteris filix-mas and Dryopteris marginalis.

Natural Order.—Filices.

Its activity is due to an *oleoresin*, which is officinal.

Dose, fʒss–j of oleoresin, taken in the morning after fasting twenty-four hours, and repeat in three hours. Used mostly for tapeworm.

Officinal Name, PEPO. *English Name*, PUMPKIN SEED.

Definition.—The *seed* of Cucurbita pepo.

Natural Order.—Cucurbitaceæ.

Dose, of seed (powdered with sugar), ʒj–ij.

ESSENTIALS OF MATERIA MEDICA. 345

Turpentine, in doses of f℥ss, is occasionally given with twice its bulk of castor oil, for both tape- and round worms.

Officinal Name, GRANATUM. *English Name*, POMEGRANATE.
Definition.—The *bark* of the *stem* and *root* of Punica granatum.
Natural Order.—Lythrarieæ. *Habitat.*—India. Cultivated in United States.
Contains the active alkaloids **pelletierine** and **isopelletierine.**
The decoction (℥ij–Oij of water boiled to Oj) is given in doses of f℥ij, before breakfast; repeated if necessary.
Dose of pelletierine, gr. v–x.

Officinal Name, KAMALA. *English Name*, KAMALA.
Definition.—The *glands* and *hairs* of the capsules of Mallotus philippinensis.
Natural Order.—Euphorbiaceæ. *Habitat.*—East Indies.
Contains the active resinoid *rottlerin*.
Dose, ʒj–ij of powder, given in syrup.

FAMILY III.—DIGESTANTS.

Remedies which increase the action of the gastric and intestinal juices in so far as their solvent power is concerned, and thereby render digestion easier and quicker of accomplishment.

Officinal Name, PEPSINUM SACCHARATUM. *English Name*, PEPSIN—SACCHARATED PEPSIN.

Definition.—A proteolytic ferment or enzyme from the fresh stomach of the pig. Capable of digesting not less than 3000 times its own weight of freshly coagulated and disintegrated egg-albumen. This is pure pepsin. When we add 90 per cent. of sugar of milk, we obtain saccharated pepsin. This digests 300 times its weight of albumen.

Dose, gr. v–xx.

Officinal Name, PANCREATINUM. *English Name*, PANCREATIN.

Definition.—A mixture of the enzymes found in the *pancreas* of the Mammalia, that of the hog being generally used.

Dose, gr. v–x.

EXTRACT OF MALT.*

Definition.—The *seeds* of barley (Hordeum distichum, *natural order*, Gramineæ) caused to germinate artificially and then dried. The extract should be free from starch.

Dose, f3j–f3j.

PAPAIN.*

Definition.—A ferment from the *fruit* of Carica papaya, a South American tree.

---

* Not officinal.

Natural Order.—Papayaceæ. *Habitat.*—South America.
Dose, gr. j-x.

FAMILY IV.—ABSORBENTS.

Remedies used to absorb acid and deleterious materials, offensive discharges, secretions, etc., both in the alimentary canal and externally. They are animal and vegetable charcoal.

Officinal Name, CARBO ANIMALIS. *English Name,* ANIMAL CHARCOAL.
Definition.—Charcoal prepared from bones, blood, etc.
Dose, ℨss.
Officinal Preparation.
Carbo Animalis Purificatus.

Officinal Name, CARBO LIGNI. *English Name,* CHARCOAL.
Definition.—Charcoal prepared from soft wood and very finely powdered. Used to dress foul wounds and ulcers, and to absorb noxious gases.

FAMILY V.—DISINFECTANTS.

Substances employed for the prevention of noxious miasmata or effluvia. Of course, heat,—both dry and moist,—air, water, ventilation, and proper disposal of infected excreta are of the utmost import-

ance. Besides these, the various salts of iron and lead, forming sulphurets, are considered important. The oxides of iron convert ordinary oxygen into ozone, which is itself a disinfectant.

COPPERAS—IMPURE SULPHATE OF IRON—

Is an important destructive disinfectant but is not strictly a germicide. Its sole use is to alter the course of putrefaction and destruction of the products thereof. May be used in solution, powder, or crystals, according to the mass to be acted upon. Is extremely valuable for use in cesspools, etc.

LIME

Is of use only as a destructive agent; it tends to prevent odor, but is useless in sewers, privies, etc.

CORROSIVE SUBLIMATE.—BICHLORIDE.—BICHLORIDE OF MERCURY.

Very powerful, both antiseptic and germicidal—even in weak solution. Owing to its poisonous nature care must be observed in its use.

CARBOLIC ACID.

An active germicide. From it we obtain creosols, cresylic acid, creolin, lysol, etc., all also germicides.

Creolin is an emulsion of creosol obtained by means of resin soap.

Lysol is said to contain 50 per cent. of creosol. Both creolin and lysol will mix with water, alcohol, and ether.

PERMANGANATE OF POTASSIUM.

A valuable disinfectant and germicide, but of limited power as it yields up its own oxygen and becomes inert. It will destroy most, if not all, alkaloids in a sufficient length of time, and has been used with success in morphine poisoning, given in doses one-third larger than the amount of morphine ingested.

CHLORINE.

The gas is a powerful germicide, but is exceedingly dangerous, and is apt to injure the clothing as well as the wearer thereof.

Officinal Preparation.

Aqua Chlori contains at least 0.4 per cent. of chlorine gas.

Diluted, may be used as a gargle in diphtheria and as a stimulant in the washing of foul ulcers.

Officinal Name, CALX CHLORATA. *English Name,* CHLORINATED LIME.

Definition.—Composed of calcium hypochlorite and calcium chloride. Owes its activity to the chlorine (25 per cent.) it yields when exposed to the air. Used as a disinfectant only.

ESSENTIALS OF MATERIA MEDICA. 355

Officinal Name, LIQUOR SODÆ CHLORATÆ. *English Name*, LABARRAQUE'S SOLUTION.

Definition.—Solution of chlorinated soda. Contains hypochlorite of soda, and may be used in the same manner as chlorinated lime, or properly diluted in the same way as aqua chlori.

Officinal Name,   *English Name*,
ACIDUM BORICUM.   BORIC OR BORACIC ACID.
SODII BORAS.   BORAX.

Definition.—Used as a dusting powder and in solution for wounds, ulcers, abscesses, burns, etc. Is also efficient in neutralizing ammoniacal urine and in cystitis due thereto, and as an eyewash.

Dose, acid, gr. v-x ; salt, gr. xx.

Officinal Name, ACIDUM SULPHUROSUM. *English Name*, SULPHUROUS ACID.

Definition.—This acid and its salts are very efficient in destroying the low forms of life connected with fermentation and putrefaction, and for this reason form an excellent preservative of organic matter.

Officinal Name, NAPHTALINUM. *English Name*, NAPHTALENE.

Definition.—A hydrocarbon obtained from coal-tar. Poisonous to the lower forms of life. Has supplanted camphor as a destroyer of moths. Also used

externally as an antiseptic dressing and in certain parasitic skin diseases, but beta-naphtol is superior to it, and oftener used.

Dose, gr. ij–viij in capsule.

Officinal Name, NAPHTOL. *English Name*, BETA-NAPHTOL.

Definition.—Prepared by heating naphtalin with sulphuric acid, then fusing with alkaline hydrates.

Alpha-naphtol is not officinal, but beta-naphtol is.

The odor faintly suggests carbolic acid; occurs in pale buff or colorless crystals, freely soluble in alcohol, slightly in water. May be used internally as an antiseptic, and externally for the same indications as naphtalene.

Officinal Name, AQUA HYDROGENII DIOXIDI. *English Name*, HYDROGEN DIOXIDE OR PEROXIDE.

H_2O_2, water, plus one atom of oxygen, is a powerful germicide; effervesces when brought into contact with pus, destroying the pus-corpuscles; is a powerful deodorant and an exceedingly valuable local antiseptic. Generally used in diluted solution. Probably not poisonous, but as a matter of precaution the solution used should not be stronger than the officinal solution,—three per cent., by weight, of the pure dioxide,—equal to about ten volumes of available oxygen.

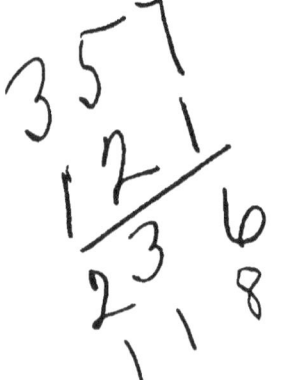

CHAPTER IV.

POISONING.

Antidotes and Treatment.

In all cases, except those specially mentioned, the stomach should be at once evacuated either by (1) emetic (mustard, salt-water, etc.), (2) hypodermic injection of apomorphinæ, or (3) the stomach-pump.

It should be borne in mind that **tannic acid** is an antidote for all the alkaloids and such vegetable drugs as depend for their activity on the alkaloids they contain. It is also the antidote for antimony, and can do no harm even in large amount.

ÆTHER. ETHER.

Stop the inhalation; artificial respiration, atropine and strychnine hypodermically, fresh air, alternate hot and cold douche, electricity.

ALCOHOL (Acute).

Evacuate the stomach, hot and cold douche, electricity, strychnine hypodermically.

ALKALIES: LIME. POTASSA. SODA.

Weak acid, administered freely and at once. (Vinegar is, as a rule, easily and quickly obtained.)

AMMONIUM. AMMONIA.

Neutralize by acid (dilute acetic acid—vinegar—is nearly always at hand); use oils and demulcents to protect the irritated surfaces; opium if necessary. Edema of the glottis or larynx demands tracheotomy.

ANTIMONII ET POTASSII TARTRAS. TARTAR EMETIC.

Evacuate the stomach; give the chemic antidote, tannic acid; opium if needed, and stimulants if required.

ARGENTI NITRAS. NITRATE OF SILVER.

Common salt (sodium chloride) forms the insoluble chloride of silver, which is inert. Treat symptoms as they arise.

ATROPINE. BELLADONNA.

Evacuate stomach (emetics), bowels (cathartics), and bladder (catheter) to prevent further absorption. Tannic acid should be given in large doses. Use morphine hypodermically as a physiologic antidote.

ACIDUM ARSENOSUM. ARSENIC OR ANY OF ITS PREPARATIONS.

Evacuate the stomach. Give the *fresh* antidote (hydrated oxide of iron with magnesia) in large doses.

ACIDUM CARBOLICUM. CARBOLIC ACID.

Very rapid in its action. Evacuation of the stomach is useless; use the antidote, a **soluble sulphate**, —preferably magnesia sulphate,—in large amounts, during all stages, as it is capable of neutralizing the poison even after absorption, provided the length of time it has been absorbed has not been too great.

ACIDS.

HYDROCHLORIC OR MURIATIC. NITRIC. SULPHURIC.

First, give alkalies or alkaline earths; soap or *very weak* ammonia solution, to neutralize the acid; then give demulcents, oils, albumens, etc., to protect the injured surfaces and treat the symptoms as they arise.

Hydrochloric acid causes a slight blistering and yellow tinge to surfaces with which it is brought into contact.

Nitric acid causes a persistent orange-yellow stain.

Sulphuric acid blackens and chars the tissues with which it is brought into contact.

ACIDUM HYDROCYANICUM.

An overdose kills almost immediately; treatment is generally useless, but if possible evacuate the stomach; atropine and strychnine hypodermically. Ammonia by the mouth, inhalation, and intraven-

ously. Hot and cold douche, electricity, and artificial respiration.

ACIDUM OXALICUM.

Lime, chalk (calcium carbonate), present in every house as tooth powder, forms an insoluble oxalate and is the *only* antidote. Must be given immediately. If necessary, use the lime from wall or fence.

ACONITUM. ACONITE.

Evacuate the stomach. Keep the head at the same level as the rest of the body. Give alcohol, ether, ammonia, digitalis, hypodermically, if necessary.

CHLORAL. CHLORAL HYDRATE.

Keep up the temperature. Stimulate the heart, use electricity, artificial respiration, and use atropine as required. Chloral is a good antidote in strychnia poisoning, but strychnia is of no use in chloral poisoning.

CHLOROFORM.

Stop the inhalation; artificial respiration, electric battery, atropine, strychnine, etc.

COLCHICUM.

Give emetics and tannic acid. Use opium and stimulants.

ESSENTIALS OF MATERIA MEDICA. 367

COPPER ACETATE. CUPRUM ACETAS.
COPPER SULPHATE. CUPRUM SULPHAS.
COPPER ACETATE (IMPURE). VERDIGRIS.

Potassium ferrocyanide forms an insoluble compound with the above. Use demulcents, opium to allay pain, and treat the symptoms as they arise.

In chronic poisoning, use iodide of potassium, gr. x, t. i. d. for continued periods of time.

CYANIDE OF POTASSIUM.
CYANIDE OF SILVER.

Treat poisoning the same as hydrocyanic acid.

DIGITALIS. FOXGLOVE.

Use emetics and cathartics. Tannic acid is unreliable but is sometimes used. Give the physiological antidotes alcohol, opium, and ammonia.

HYDRARGYRUM. MERCURY AND ITS PREPARATIONS.

Albumen (white of egg, milk, or wheat flour) followed by an emetic; demulcents; opium if required.

IODUM. IODINE.

Evacuate the stomach and give liquids containing *starch* in large amount.

MORPHINE.

Tannic acid; evacuate the stomach; give strong black coffee; keep the patient awake; use electricity, hot and cold douche. Use atropine to its limit; artificial respiration, long continued if necessary. Strychnia is of value in large doses.

NUX VOMICA. STRYCHNINA.

Evacuate the stomach; give tannic acid as chemic antidote. Use chloral and potassium bromide in large doses pushed to effect. Produce free catharsis and diuresis; use the catheter if necessary.

OPIUM. (See Morphine.)

PHOSPHORUS.

French oil of turpentine, which has been successfully used abroad, is practically unobtainable in America. Sulphate of copper forms an insoluble sulphate, and therefore is the antidote most frequently employed in this country.

Give gr. x doses and repeat.

PLUMBUM. LEAD.

Evacuate the stomach. Give a soluble sulphate or dilute sulphuric acid (forming an insoluble sulphate), demulcents if necessary, and use morphine for pain and vomiting.

In chronic lead poisoning, dilute sulphuric acid as a habitual drink, magnesium sulphate as both purge and antidote, and potassium iodide in gr. x doses continued for some weeks will be found of valuable service. Alum is sometimes, though rarely, used.

VERATRUM VIRIDE.

An overdose generally produces emesis and so renders poisoning rare. Treat symptomatically. Give heart stimulants as required.

ZINCI SULPHAS.

Alkalies and alkaline carbonates produce insoluble precipitates.

Give demulcents and treat symptomatically.

ADDENDA.

In order to facilitate the study of the natural orders, the following table is here included:

Algæ, ... Chondrus.
Anacardieæ, ... Rhus glabra.
Apocynaceæ, ... { Aspidosperma.
                     { Strophanthus.
Aristolochiaceæ, ... Serpentaria.
Aroideæ, ... Calamus.
Berberideæ, ... Podophyllum.
Burseraceæ, ... Myrrha.
Celastrineæ, ... Euonymus.
Chenopodiaceæ, ... Chenopodium.
Coleoptera, ... Cantharis.

Compositæ, ... {
  Anacyclus pyrethrum.
  Arnicæ.
  Artemisa pauciflora (santonin).
  Anthemis.
  Erigeron.
  Eupatorium.
  Grindelia.
  Lactucarium.
  Tanacetum.
  Taraxacum.
}

ADDENDA.

Coniferæ,	{ Abies canadensis. / Abies excelsa. / Juniperus communis. / Juniperus sabina. / Pinus palustris.
Convolvulaceæ,	{ Jalapa. / Scammonium.
Cruciferæ,	Sinapis alba and sinapis nigra.
Cucurbitaceæ,	{ Colocynthis. / Elaterium. / Pepo.
Cupuliferæ,	{ Quercus alba. / Quercus lusitanica. / Fagus sylvatica (beech).
Ericaceæ,	{ Chimaphila. / Uva Ursi.
Euphorbiaceæ,	{ Cascarilla. / Kamala. / Oleum ricini. / Oleum tiglii.
Filices,	Aspidium.
Fungi,	{ Ergota. / Ustilago.
Gentianeæ,	{ Gentiana. / Chirata.
Geraniaceæ,	Geranium.
Gramineæ,	{ Rye (secale). / Zea mays.
Guttiferæ,	Cambogia.
Juglandaceæ,	Juglans.
Labiatæ,	{ Hedeoma. / Melissa. / Mentha piperita. / Mentha viridis. / Salvia.

ADDENDA. 377

Laurineæ, ... {
Camphora.
Cinnamomum zeylanicum.
Sassafras variifolii.

Leguminosæ, ... {
Acacia.
Balsamum Peruvianum.
Balsamum tolutanum.
Cassia fistula.
Catechu.
Copaiba Langsdorffii.
Glycyrrhiza.
Hæmatoxylon.
Kino.
Physostigma.
Scoparius (sparteine).
Senna.
Tamarindus.
Tragacantha.

Lichenes, ... Cetraria.

Liliaceæ, ... {
Aloes.
Allium.
Colchicum.
Convallaria majalis.
Sarsaparilla.
Scilla.
Veratrina.
Veratrum viride.

Lineæ, ... {
Coca.
Linum.

Lobeliaceæ, ... Lobelia.

Loganiaceæ, ... {
Gelsemium.
Ignatia.
Nux vomica.
Spigelia.

25

ADDENDA.

Lythrarieæ,	Punica granatum.
Malvaceæ,	{ Althæa. Gossypii.
Menispermaceæ,	{ Calumba. Pareira.
Myristicaceæ,	Myristica.
Myrtaceæ,	{ Caryophyllus. Eucalyptus. Oleum cajuputi. Pimenta.
Oleaceæ,	Manna.
Papaveraceæ,	{ Opium. Sanguinaria.
Papayaceæ,	Papain.
Piperaceæ,	{ Piper. Matico. Cubeba.
Polygaleæ,	{ Krameria. Senega.
Polygonaceæ,	Rheum.
Ranunculaceæ,	{ Aconitum. Adonidine. Cimicifuga. Hydrastis.
Rhamnaceæ,	{ Rhamnus frangula. Rhamnus purshiana.
Rosaceæ,	{ Brayera (cusso). Prunus virginiana. Amygdala. Rosa.
Rubiaceæ,	{ Caffea arabica. Cinchona. Ipecacuanha.

ADDENDA. 381

Ruminantia,	Musk.
Rutaceæ,	Aurantii. Buchu. Pilocarpus. Ruta.
Scitamineæ,	Cardamomum. Zingiber.
Santalaceæ,	Oleum santali.
Sapindaceæ,	Paullinia sorbilis (guarana).
Sterculiaceæ,	Theobroma.
Styraceæ,	Benzoinum.
Solanaceæ,	Belladonna. Capsicum. Hyoscyamus. Stramonium. Tabacum.
Thymelæaceæ,	Mezereum.
Ternstrœmiaceæ,	Camellia thea (Thea sinensis .
Urticaceæ,	Cannabis indica. Humulus. Ulmus.
Umbelliferæ,	Ammoniacum. Anisum. Asafœtida. Carum. Conium. Coriandrum. Fœniculum. Petroselinum sativum.
Valerianeæ,	Valeriana.
Zygophylleæ,	Guaiaci.

INDEX.

	PAGE		PAGE
ABBREVIATIONS,	61-66	Apomorphine,	263
Absorbents,	349	Apothecaries' measure,	17
Acacia,	325	weight,	17
Aceta, list of,	97	Aquæ, list of,	91
Acetanilid,	237	Aristol,	217
Acetic acid,	181	Arnica,	177
Aconite,	177	Aromatics,	243
Active principles,	119	Argentum,	199
Adeps,	333	Arrowroot,	331
Adonidine,	173	Arsenic,	211
Administration,	71	Asafœtida,	131
Adjuvant,	13	Aspidosperma,	163
Æther,	137	Aspidium,	343
Alcohol,	167	Astringents,	125, 183
Alkaloids, doses of,	119	Atropine,	151
definition of,	119	Aurantii cortex,	253
Allium,	309	flores,	255
Allspice,	247	Auri et sodii chloridum,	215
Aloes,	275	Average doses,	117
Alteratives,	209, 125	Azedarach.	341
Altheæ,	329		
Alum,	193		
American worm-seed,	341	BALM,	259
Ammonia,	165	Balsam of Peru,	307
Ammoniacum,	305	of Tolu,	307
Ammonium chloride,	305	Barley,	331
Amyl nitrite,	159	Basham's mixture,	203
Anesthetics,	123, 137	Basis,	13
Anise,	253	Bearberry,	293
Antacids,	337	Belladonna,	149
Anthelmintics,	339	Benzoic acid,	235
Anthemis,	259	Benzoinum,	235, 307
Antidotes,	183, 359	Beta-naphtol,	357
Antifebrin,	237	Bismuth,	195
Antimony,	173	Black cohosh,	135
Antiperiodics,	125	Black snakeroot,	135
Antipyretics,	125	Blatta,	287
Antipyrin,	235	Bloodroot,	263
Antispasmodics,	123	Blue mass,	211
Apiol,	313	Blue ointment,	211

INDEX.

	PAGE
Boneset,	241
Borax,	355
Boric acid,	355
Bougia,	115
Brimstone,	271
Bromide of potassium,	157
of ammonium,	157
of lithium,	157
of sodium,	157
Bromine,	323
Bromoform,	139
Broom,	287
Brucine,	155
Buckthorn,	269
Buchu,	293
Butternut,	275
CACAO BUTTER,	333
Cachets,	115
Cachous,	115
Caffeine,	169
Calamus,	259
Calcium,	339
Calisaya bark,	223
Calomel,	211
Calumba,	241
Cambogia,	283
Camphor,	131
Cannabis indica,	149
Cantharides,	319
Capsicum,	247
Caraway,	251
Carbolic acid,	229, 363
Cardamon,	247
Cardiac depressants,	125
stimulants,	125
Cardiants,	123
Carum,	251
Caryophyllus,	245
Cascara sagrada,	269
Cascarilla,	261
Cassia fistula,	267
Castor oil,	271
Catechu,	187
Cathartics,	265
Cera,	333
Cerates,	105
Cerii oxalas,	197
Cetaceum,	333
Cetraria,	327
Chamomile,	259
Charcoal,	349
Chartæ,	113
Chenopodium,	341

	PAGE
Chimaphila,	295
Chirata,	243
Chondrus,	327
Chloral,	145
Chromic acid,	323
Chlorinated lime,	353
Chlorine,	353
Chloroform,	139
Cimicifuga,	135
Cinchona,	223
Cinchonine,	225
test for,	227
Cinnamon,	243
Citric acid,	179
Clark's rule,	73
Classification of drugs,	121
Cloves,	245
Coca,	153
Cocaine,	141, 153
Codeine,	145
Cod-liver oil,	217
Colchicum,	219
Collodion,	335, 115
Colocynth,	281
Combination of medicines,	77
Compound cathartic pill,	281
prescriptions,	11
spirits of ether,	133
Confections,	103
Conium,	163
Convallaria,	171
Copaiba,	299
Copper,	199
Copperas,	351
Coriander,	251
Corn ergot,	319
Corrective,	13
Corrigent,	13
Corrosive sublimate,	351
Cotton root bark,	317
Cowling's rule,	73
Cranesbill,	191
Cream of tartar,	289
Creosote,	231
Croton oil,	285
Cubebs,	299
Cusso,	341
DANDELION,	223
Daphnin,	221
Daturine,	151
Decoctions, doses of,	119
list of,	89
Delirifacients,	123

INDEX.

	PAGE
Demulcents,	375
De Young's rule,	73
Depresso-motors,	125
Dewees' carminative,	131
Diaphoretics,	301
Digestants,	345
Digitalis,	169
Diluents,	335
Directions to druggist,	61–66
Disinfectants,	349
Dispensatory,	87
Diuretics,	285
Domestic measures,	19
Donovan's solution,	211
Doses, average,	117
calculation of,	33–37
Dover's powder,	143
Dragées,	115
Drastics,	279
Drops, size of,	21

	PAGE
ELATERIN,	283
Elixirs, list of,	101
Emetics,	261–265
Emetine,	261
Emmenagogues,	311
Emollients,	331
Emulsions, dose of,	117
list of,	95
Enema,	115
English measure,	17
weights,	17
Epispastics,	319
Epsom salt,	271
Ergot,	315
Errhines,	319
Escharotics,	323
Eserine,	157
Ether,	137
Ethyl chloride,	141
Eucaine,	141
Eucalyptus,	227
Euonymus,	269
Eupatorium,	241
Excipients,	13
Excito-motors,	123
Expectorants,	303
Extracts, doses of,	117
list of,	105
Extraneous remedies,	127

	PAGE
FENNEL,	249
Flaxseed,	329

	PAGE
Fluid extracts, dose of,	117
list of,	107–109
Fœniculum,	249
Fowler's solution,	211
Frangula,	267

	PAGE
GALLA,	185
Gallic acid,	185
Gamboge,	283
Garlic,	309
Gelsemium,	161
General remedies,	123
Gentian,	239
Geranium,	191
Ginger,	249
Glauber's salt,	277
Glonoinum,	159
Glucoside,	119
Glycerin,	331
Glycerites,	99
Glycyrrhiza,	327
Gold and sodium chloride,	215
Golden seal,	239
Gossypii radicis cortex,	317
Grammar,	43–59
Granatum,	345
Granulum,	115
Grindelia,	305
Guaiac,	221
Gum Arabic,	325

	PAGE
HÆMATOXYLON,	189
Hedeoma,	315
Hemlock,	163
Henbane,	153
Hive syrup,	287
Hoffmann's anodyne,	133
Homatropine,	151
Hops,	135
Hordeum,	331
Humulus,	135
Hydrargyrum,	211
Hydrastis,	239
Hydrobromic acid,	157, 161
Hydrochloric acid,	207
Hydrocyanic acid,	179
Hydrogen dioxide,	357
Hyoscine,	147, 153
Hyoscyamus,	153

	PAGE
ICELAND MOSS,	327
Ichthyol,	223

INDEX.

	PAGE
Ignatia,	155
Incompatibles,	77
Infusions, doses of,	89
list of,	119
Inscription,	11
Iodine,	215
Iodoform,	217
Iodol,	217
Ipecac,	261
Irish moss,	327
Iron,	201
JABORANDI,	301
Jalap,	279
James' powders,	175
Jamestown weed,	151
Juglans,	275
Juniper,	295
KAMALA,	345
Kino,	187
Krameria,	189
LABARRAQUE'S SOLUTION,	355
Lactic acid,	209
Lactucarium,	135
Lady Webster's pill,	275
Lanolin,	331
Lard,	333
Laughing gas,	137
Laudanum,	141
Laxatives,	265
Lavender,	255
Lead,	193
Lettuce opium,	135
Levant worm seed,	341
Lime,	351
Liniments,	113
Linum,	329
Liquor ammonii acetatis,	303
gutta-percha,	337
Liquores, list of,	89
Liquorice,	327
Lithium,	291
Lobelia,	159
Logwood,	189
Local remedies,	125–127
Lunar caustic,	199
Lupulinum,	135
Lugol's solution,	215

	PAGE
MAGNESIA,	269
Male fern,	343
Malt,	347
Manganese,	205
Manna,	267
Maranta,	331
Marshmallow,	329
Materia medica,	85
Matico,	301
May apple,	283
Melissa,	259
Mellita,	99
Mentha piperita,	257
viridis,	257
Menthol,	231
Mercury,	211
Metric system,	23–27
Mezereum,	221
Mistura, doses of,	119
list of,	95
Monsel's solution,	203
Morphine,	145
Moschus,	129
Mucilagines,	95
Musk,	129
Mustard,	321
Myristica,	245
NAPHTALINE,	355
Naphtol,	357
Narceine,	141
Narcotine,	141
Natural orders, list of,	373–381
Nervines,	123, 129
Nicotine,	161
Nitric acid,	207
Nitrite of amyl,	159
Nitroglycerin,	159
Nitrohydrochloric acid,	207
Nitrogen monoxide,	137
Nitrous oxide,	137
Norwood's tincture,	175
Numerals,	67
Nutmeg,	245
Nutrients,	123
Nux vomica,	155
OIL OF CAJUPUT,	251
erigeron,	295
sandalwood,	297
sassafras,	253
vitriol,	205
wintergreen,	233

INDEX.

Oils, doses of,	119
list of,	99
test for,	101
Ointments,	105
Oleoresinæ, list of,	101
Oleum cajuputi,	251
erigerontis,	295
gaultheriæ,	233
morrhuæ,	217
ricini,	271
santali,	297
sassafras,	253
theobromatis,	333
tiglii,	285
Opium,	141
poisoning,	369
Orange flowers,	255
peel,	253
Oxalate of cerium,	197
Oxalic acid,	181
Oxytocics,	315
PANCREATIN,	347
Papain,	347
Papers,	113
Paraldehyde,	147
Paregoric,	141
Pareira,	293
Pellitory,	319
Pennyroyal,	315
Pepo,	343
Pepper (black),	247
(red),	247
Peppermint,	257
Pepsin,	347
Pepsinum saccharatum,	347
Permanganate of potassium,	353
Pesoaria,	115
Petrolatum,	335
Pharmacology,	85
Pharmacopœia,	87
Pharmacy,	87
Phenacetine,	237
Phenic acid,	229
Phenylic acid,	229
Phosphoric acid,	217
Phosphorus,	209, 369
Physiologic action,	85
Physostigma,	157
Picric acid,	229
Pills,	111
Pilocarpus,	301
Pilulæ,	111
Pimenta,	247

Pinkroot,	339
Piper,	247
Piperazine,	291
Piperine,	247
Pipsissewa,	295
Pix burgundica,	321
Pix liquida,	311
Plasters,	111
Plumbum,	193
Plummer's pills,	175
Podophyllum,	283
Poisons,	359, 375
Pomegranate,	345
Potassii et sodii tartras,	279
Potassium,	289
permanganate,	353
Powders,	113
Prescriptions,	9-11
Protectives,	335
Prunus Virginiana,	241
Pulveres,	113
Pulvis effervescens compositus,	279
Pumpkin seed,	343
Purges,	271
Pyrethrum,	319
QUASSIA,	237
Quercus alba,	189
Quinidine,	225
tests for,	227
Quinine,	225
tests for,	227
RED PEPPER,	247
Resinæ,	103
Resorcin,	233
Rhamus frangula,	269
purshiana,	267
Rhatany,	189
Rheum,	273
Rhubarb,	273
Rhus glabra,	193
Rochelle salt,	279
Roses,	189, 191
Rosemary,	257
Rubefacients,	321
Rue,	313
Ruta,	313
SABINE,	313
Sage,	255

INDEX.

	PAGE
Sago,	331
Salicin,	235
Salicylic acid,	233
Salol,	235
Salvia,	255
Sanguinaria,	263
Santonica,	341
Santonin,	341
Sarsaparilla,	219
Sassafras,	221
medulla,	329
pith,	329
Savine,	313
Scammony,	281
Scilla,	285
Scoparius,	287
Seidlitz powder,	279
Senega,	305
Senna,	277
Serpentaria,	259
Sialagogues,	319
Signature,	11
Silver,	199
Simple bitters,	237
Sinapis alba,	321
nigra,	321
Slippery elm,	325
Sodii boras,	355
phosphas,	277
sulphas,	277
Sodium,	337
Solutions,	89
Somnifacients,	123
Spanish flies,	319
Sparteine,	173
Spearmint,	257
Spermaceti,	333
Spigelia,	339
Spirit of Mindererus,	165, 303
of nitroglycerin,	159
Spirits,	91
Spiritus ætheris compositus,	133
nitrosi,	289, 303
frumenti,	167
glonoini,	159
odoratus,	169
vini gallici,	167
Squill,	285
Stomachics,	237
Stramonium,	151
Strontium,	293
Strophanthus,	171
Strychnine,	155
test for,	155
Subscription,	11

	PAGE
Sulphate of copper,	191
of zinc,	197
Sulphonal,	147
Sulphur,	271
Sulphuric acid,	205
Sulphurous acid,	335
Sumach,	193
Superscription,	11
Suppositories,	103
Sweet flag,	259
Sweet spirits of nitre,	289
Syrups, list of,	97
doses of,	119
TABACUM,	161
Tamarinds,	267
Tanacetum,	315
Tannic acid,	183
Tansy,	315
Tapioca,	331
Tar,	311
Taraxacum,	223
Tartar emetic,	173
Tartaric acid,	179
Terebene,	311
Tetronal,	147
Terebintha,	297
canadensis,	299
Therapeutics,	87
Thoroughwort,	241
Theobromine,	287
Thymol,	231
Tinctures, list of,	91-93
doses of,	117
Tobacco,	161
Tonics,	125, 201
Traumaticine,	337
Tragacanth,	325
Trional,	147
Troches,	103
Tropacocaine,	153
Tully's powder,	145
Turpentine,	297, 345
Turpeth mineral,	213
ULMUS,	325
Unguenta,	105
Ustilago,	317
Uva ursi,	293
VALERIAN,	129
Vallet's mass,	203

INDEX.

	PAGE		PAGE
Vaseline,	335	Weights and measures,	23-37
Vegetable acids,	179	Whisky,	167
Vehicle,	13	White oak,	189
Veratrina,	177	White wax,	333
Veratroidine,	175	White wine,	167
Veratrum viride,	175	Wild-cherry bark,	241
Vina,	99	Wine measure,	19
Vinegars,	181	Wines,	99
Virginia snake root,	259		
Vitriol, blue,	199		
green,	203	**YELLOW CINCHONA**,	223
oil of,	205	Yellow jasmine,	161
white,	197	Yellow wax,	333
WATERS,	91	**ZEA MAYS**,	297
Wahoo,	263	Zinc,	197
Warburg's tincture,	227	Zinziber,	249

www.ingramcontent.com/pod-product-compliance
Lightning Source LLC
Chambersburg PA
CBHW032021220426
43664CB00006B/323